'Surviving a death related to drug or alcohol abuse can be a particularly tragic experience. This book uniquely deals with the process of bereavement in these special circumstances, and unveils a world still unknown to many.'
—*Professor Diego De Leo AO, DSc, Director, Australian Institute for Suicide Research and Prevention, Griffith University, Brisbane, Australia*

'Any society that claims to be a caring one should recognise and respond compassionately to the doubly painful bereavement experiences that this group of researchers discuss. The story they tell is a moving one, but it is not always reassuring. It seems that we have been ignoring what is in fact a common problem – losing a loved one who had an alcohol or drug problem – and that those of us who meet such bereaved family members have not always been as caring as we should have been. There are important lessons to be learned here.'
—*Jim Orford, Emeritus Professor of Clinical and Community Psychology, University of Birmingham, Birmingham, England*

Families Bereaved by Alcohol or Drugs

Individuals bereaved by the drug- or alcohol-related death of a family member represent a sizeable group worldwide. *Families Bereaved by Alcohol or Drugs* is the long-awaited result of an important and ambitious research project into the experiences commonly encountered by members of this stigmatised and vulnerable group.

Based on focus groups with the practitioners and service personnel who support grieving relatives following the loss of a loved one to alcohol or drugs, as well as interviews with the largest qualitative sample of adults bereaved by substance use that has been reported to date, this much-needed contribution to research on addiction and bereavement identifies four major reasons why grief following this tragic kind of death is particularly difficult. By examining the experiences of a wide range of stakeholders, including practitioners and policymakers in health, social care and the criminal justice system, the research contained within this book underscores the large number of organizations that play a role in the implementation of official procedure following a drug- or alcohol-related death and identifies significant gaps in the system that bereaved individuals must negotiate.

Grounded in extensive and rigorous academic research, *Families Bereaved by Alcohol or Drugs* is essential reading for academics, researchers and postgraduate students in the fields of mental health and addiction, social work and social studies, psychology, family studies and bereavement. The book should also be of interest to anyone with a professional interest in bereavement or substance use.

Dr Christine Valentine is a Research Fellow and member of the Centre for Death and Society at the University of Bath. She is a founding member of the Association for the Study of Death and Society. Christine has published widely on the social and cultural shaping of bereavement in Britain and Japan, on funeral welfare systems for people on low income both nationally and internationally, and on funeral directing in the 21st century. She was the lead researcher for the ESRC-funded research on which the present book is based.

Explorations in Mental Health

For a full list of titles in this series, please visit www.routledge.com

Families Bereaved by Alcohol or Drugs

Research on Experiences, Coping and Support

Edited by
Christine Valentine

Routledge
Taylor & Francis Group

LONDON AND NEW YORK

First published 2018
by Routledge

2 Park Square, Milton Park, Abingdon, Oxfordshire OX14 4RN
52 Vanderbilt Avenue, New York, NY 10017

Routledge is an imprint of the Taylor & Francis Group, an informa business

First issued in paperback 2018

British Library Cataloguing-in-Publication Data
A catalogue record for this book is available from the British Library

Library of Congress Cataloging-in-Publication Data
A catalog record for this book has been requested

ISBN: 978-1-138-94708-5 (hbk)
ISBN: 978-0-367-17865-9 (pbk)

Typeset in Bembo
by Apex CoVantage, LLC

We dedicate this book to Joan Hollywood, a member of the research team that carried out the research on which this book is based. Joan was the inspiration behind this first major qualitative research project to better understand and improve support for adults bereaved after a substance-related death; without her, the project would never have happened. Following her son's death in 2008 from drug and alcohol use, being unable to find support for grieving a substance-related death, Joan, with her husband, Paul, founded the support organisation, Bereavement Through Addiction (BTA), in Bristol. Since 2012 BTA has provided a helpline, support groups and an annual memorial service for others bereaved in this way, as well as training to other organisations. Joan became a tireless campaigner for people bereaved by substance use. Through BTA she developed an extensive network of bereaved people and practitioners involved with drug and alcohol treatment, related deaths and bereavement support. Joan died on March 10, 2015, before the project was completed.

Contents

Figures and tables

Figures

Tables

Foreword

It is both an honour and a privilege to have this special opportunity to introduce this important new book on drug death bereavement, an essential subject that has surprisingly eluded researchers' interests. In many ways, this book could not be more timely and path-breaking.

First, *Families Bereaved by Alcohol or Drugs: Experiences, Coping and Support* offers the first-ever research monograph on this beleaguered and under-served population. Second, within these pages, we listen to the voices of the substance-death bereaved themselves. They share their anguish and sadness, recount their loss stories and offer important details of the stigma and blame induction that society exacts upon the close relatives of those succumbing to deaths from drugs or alcohol. We learn that drug death bereaved families sometimes endure long and painful trials before the deaths, feeling helpless as they find themselves powerless to avert the ultimate tragic outcomes of their loved ones' deaths. We learn that other family members and friends sometimes inadvertently inflict deep pain and hurt by unthinkingly expressing prejudice and antipathy toward drug dependent individuals and deny those bereaved any legitimacy to their grief. We also learn that caregivers, police, emergency personnel, healing professionals and other officials may inflict further suffering upon the bereaved with their occasionally callous and insensitive remarks.

Third, the authors have also conducted focus group interviews with a wide range of practitioners to help them gain a better understanding of the grief and bereavement needs of those affected by substance-related deaths. Then, in a final section, the authors attempt to extrapolate from all preceding chapters a set of meaningful and empirically based guidelines for practitioners to follow to better serve the bereavement needs of these mourners. Thus, the aim of this work is dedicated toward improving the quality of care available for the neglected population of substance death mourners. As I see it, this important book has been long overdue.

Taking stock of the drug death grief problem in U.S.-based research, I find much confirmation of the results obtained from this very able group of British researchers who extensively interviewed 106 substance death bereaved from England and Scotland. I will now offer a few more convergences and contrasts, paralleling the findings presented in this volume.

Presently, in the United States, the number of drug deaths has been steadily rising over the past two decades to a point where it now eclipses the peak numbers of AIDS/HIV deaths that soared to 42,000 in the mid-1990s (*New York Times*, Jan. 16, 2016). The Centers for Disease Control (CDC) tabulated a steep 137% rise in the number of drug overdose deaths since 2000, bringing the drug overdose rate numbers to over 47,000 in 2014 (*Morbidity and Mortality Weekly Reports*, December 18, 2015). Now, U.S. drug deaths exceed the yearly number of suicides and motor vehicle accident deaths. This pattern of rising drug deaths appears to be an international phenomenon. According to the Chooper's Guide (www.choopersguide.com/article/international-overdose-statistics.html) rising numbers of drug overdose deaths, exceeding the numbers from highway accidents, have now been reported in Australia, Great Britain, Scotland and Norway. Apparently, the same forces leading to a spiking of U.S. drug death rates – permissive prescription practices making pain-killing medications more readily available and a wider distribution of cheaper and synthetic opiates – have caused these sharply rising drug death rates in many countries.

In the United States drug- and alcohol-related deaths are tabulated separately. According to the CDC, during the period from 2006 to 2010 (www.cdc.gov/alcohol/fact-sheets/alcohol-use.htm), alcoholism led to an average of over 17,000 deaths annually, comprising over 2,000 alcohol poisoning deaths (*Morbidity and Mortality Weekly Reports*, January 9, 2015), with the remainder due to a variety of other causes such as liver disease fatalities and alcohol-related traffic deaths. The CDC estimates that excessive drinking alone accounts for 1 in every 10 deaths among the American working-age population (http://dx.doi.org/10.5888/pcd11.130293), and about one of every 100 deaths over all. Thus, if we calculate anywhere from 6 to 12 collateral friends and relatives for every drug or alcohol decedent, we are dealing in the United States with a drug death mourner population growing at a rate of at least half a million people yearly. Unfortunately, precious little is known about this large and rapidly expanding bereaved population.

Our own research on U.S. drug death bereaved was based on a survey research of 575 bereaved parents: 48 of whom lost a child to drug-related causes, 462 experienced a child's suicide, 24 endured a child's death through natural causes and 37 had a child die in a motor vehicle accident, drowning or from some other unintentional cause (Feigelman et al., 2012). We asked respondents an open-ended question to summarise how most of their close family members and friends had responded after the death. We then coded these comments and counted the numbers of blaming statements expressed, where others had blamed either the child or the parent for the death. Tabulations showed approximately one half of the drug death and the suicide bereaved had been exposed to blaming statements, compared to only one or two parents of the accidental death bereaved and none of those suffering from a child's death due to natural causes. We also found that both drug death and suicide death bereaved parents were much more likely than the other two groups to have experienced higher levels of grief problems, more complicated grief and higher post-traumatic stress disorder (PTSD). Thus, we learn from various locales that both the drug death and suicide bereaved are alike in being subject to being stigmatised and belittled by a wide variety of social sources and that healing professionals, too, may be among those who sometimes act unwittingly, becoming unhelpful in rendering care to these deeply wounded bereaved.

It remains a task for future research to assess whether their loved one's unsuccessful mental health and substance use treatments may have discouraged the bereaved from seeking their own professional therapeutic help. Just recently a drug death bereaved mother shared with me her own hesitancy to use treatment professionals after losing both her son and her brother to drug overdose deaths. She said the doctors kept giving her son and brother all the painkillers they asked for without ever monitoring them.

> Right now I am determined not to take any kind of pain killers for myself – I just don't trust the medical profession. Before my son's and brother's deaths I used to have normal blood pressure; now with all this sadness and turmoil, I must take drugs for hypertension, but that's it. I won't have any more of their so-called care.

Because of their scepticism toward healing professionals, many drug death and suicide bereaved in the United States seek healing assistance from support groups. Support groups for suicide survivors are relatively abundant, with more than 300 presently functioning groups across the United States. Our own and others' research finds widespread satisfaction among the suicide bereaved in these groups. Americans who are drug death bereaved may not be as fortunate as suicide survivors, with no vast networks of support groups existing exclusively for serving their healing needs. Drug death bereaved parents often find empathic support in online bereavement groups, in a few select groups exclusively serving the drug death bereaved and in general bereavement support groups such as the Compassionate Friends, serving all subgroups of bereaved parents. Compassionate Friends groups probably provide support to the largest numbers of American drug death bereaved parents. Occasional complaints have been voiced by drug death bereaved that at Compassionate Friends groups they did not experience the same empathic support available to other parents whose children died from more 'respected deaths' such as brain tumours or drownings. Thus, reports do occasionally surface from some drug death survivors that their grief support needs have not been well served in some mixed death cause support groups, and they have needed to scramble around quite a bit to find empathic support.

Another interesting contrast between the United States and the UK for the substance use death bereaved is the availability (or the lack) of governmental support. Thanks to Britain's National Health Service, there appears to be governmental interest in maintaining the health of the British citizenry (and this usually includes mental health) and to routinely offer addiction treatment services to all. Thus every health and local authority commissioning group must commission substance use services, offered free at the point of delivery, although, as this book demonstrates, services are not always extended to families or to those bereaved through addiction. Thanks also to other governmental research funding agencies, such as the Economic and Social Research Council, who have seen important knowledge benefits emerge from funding such projects as this worthy English and Scottish study, valuable data will be obtained aimed at improving mental health services for the drug death bereaved. These are important first steps in advancing the mental health needs of the drug death bereaved.

In the United States, by contrast, because of its decentralised and privatised health care system, we see government doing little else beyond tallying the numbers of drug and alcohol decedents. In the United States, unfortunately, we still have not fully committed ourselves to helping drug dependent individuals by offering them care and mental health services instead of incarceration. In the United States still, it appears we have yet to fully grasp the urgency of the drug dependency epidemic and the need to develop social services for all afflicted parties, including the drug death bereaved.

Without any further digressions, I am pleased to guide readers forward to partake in the rich intellectual dividends available from this new, trail-blazing work. Let us hope it helps to usher in a new era where the bereavement and mental health needs of the substance death bereaved become more widely understood and a new era also emerges where better mental health services become available for this presently under-served population.

William Feigelman, PhD
Emeritus Professor of Sociology
Nassau Community College
New York

References

Feigelman, W., Jordan, J. R., McIntosh, J. L., and Feigelman, B. (2012). Drug overdose death and survivor's grief. In William Feigelman, John R. Jordan, John L. McKintosh and Beverly Feigelman, *Devastating Losses: How Parents Cope With a Child's Death From Suicide or Drugs* . . . New York: Springer, Ch. 3, pp. 59–79.

Kanny, D., Brewer, R. D., Mesnick, J. B., Paulozzi, L. J., Naimi, T. S., and Lu, H. (2015). Vital signs: Alcohol poisoning deaths – United States, 2010–2012. *MMWR Morb Mortal Wkly Rep*, Jan 9, 63(53), 1238–1242.

Rudd, R. A., Aleshire, N., Zibbell, J. E., and Gladden, R. M. (2015). Increases in drug and opioid overdose deaths, United States. 2000–2014. *Mortality and Morbidity Weekly Report*, Dec 18, 64 (50), 1378–1382.

Stahre, M., Roeber, J., Kanny, D., Brewer, R. D., and Zhang, X. (2014). Contribution of excessive alcohol consumption to deaths and years of life lost in the United States. *Preventing Chronic Disease*, June 26, 11, 130293.

Contributors

Editor/Author

Christine Valentine is a Research Fellow and member of the Centre for Death and Society at the University of Bath. She is a founding member of the Association for the Study of Death and Society. Christine has researched and published widely on the social and cultural shaping of bereavement in Britain and Japan, on funeral welfare systems for people on low income both nationally and internationally and on funeral directing in the 21st century. The findings from these studies have been published in various articles, edited collections and in the book *Bereavement Narratives: Continuing Bonds in the 21st Century* (published by Routledge, 2008). She was the lead researcher for the ESRC-funded research on which this book is based.

Authors

Linda Bauld is Professor of Health Policy and Director of the Institute for Social Marketing at the University of Stirling. She also Deputy Director of the UK Centre for Tobacco and Alcohol Studies (a UKCRC Centre for Public Health Excellence covering 13 universities) and holds the CRUK/BUPA Chair in Behavioural Research for Cancer Prevention at Cancer Research UK. Linda's research focuses on the evaluation of complex interventions to improve health. Her main areas of expertise are tobacco control, smoking cessation and drug and alcohol use. She is the author of more than 100 publications in peer-reviewed journals and book chapters and has authored or edited six books. She was a co-investigator for the ESRC-funded research on which the book is based.

Peter Cartwright is a counsellor and trainer and group facilitator. Since 1998 he has worked for Adfam providing one-to-one counselling to families affected by substance use, facilitating weekly support groups, and offering helpline support and outreach work at HMP Wormwood Scrubs. Peter has contributed to the authorship of *Bereaved by Addiction* – a booklet for anyone bereaved through drug or alcohol use produced in 2013 for the charity DrugFAM. In 2015 he chaired the working group that developed practitioner guidelines, the main output from the ESRC-funded research on which this book is based. He is the primary author of those guidelines: *Bereaved Through Substance Use – Guidelines for Those Whose Work Brings Them Into Contact With Adults Bereaved After a Drug or Alcohol-Related Death*.

Allison Ford is a Research Fellow at the Institute for Social Marketing and the UK Centre for Tobacco and Alcohol Studies at the University of Stirling. She is a mixed methods researcher who has worked in public health research for over eight years. Allison has conducted and contributed to publications on a range of studies, which have included monitoring the impact of tobacco control policies on young people, youth responses to tobacco packaging design, smoking cessation in pregnancy, identifying the support needs of people bereaved through drugs or alcohol and evaluating an intervention to help people on long-term incapacity benefit return to work. She was a researcher for the ESRC-funded bereavement through substance use research on which this book is based.

Gordon Hay is Reader for the Centre for Public Health, Liverpool John Moores University. Gordon has been using quantitative and statistical methods to describe various aspects of drug or alcohol use for over 20 years. The main focus of his research has been developing and applying methods to estimate the size of hidden populations, such as opiate and/or crack cocaine users, or specific groups such as drug or alcohol users in receipt of state benefits or children affected by parental drug use. He was a consultant for the ESRC-funded bereavement through substance use research on which this book is based.

Jennifer McKell is a Research Fellow at the University of Stirling and UK Centre for Tobacco and Alcohol Studies. She is an experienced public health researcher who specialises in qualitative methods. Jennifer has conducted a range of studies on drug and alcohol use and smoking, including evaluating a range of interventions to improve health. She was joint Coordinator of the Independent Enquiry into Maximising the Recovery from Dependent Drug Use in Scotland. She has published widely on topics related to both substance use and substance use bereavement. She was a researcher for the ESRC-funded research on which this book is based.

Lorna Templeton is an Independent Research Consultant in Bristol. She has worked for over 15 years exploring how children and families are affected by the substance misuse of a relative and how research, practice and policy need to be improved to better meet the needs of this large, but often marginalised, group. Lorna is a Trustee of Adfam, a member of Alcohol Research UK's Research Panel, and a founder member of AFINet, Addiction and the Family International Network: www.afinetwork.info/. She has published widely on substance use in families and on substance use bereavement. She was a researcher for the ESRC-funded research on which this book is based.

Richard Velleman is Emeritus Professor of Mental Health Research at the University of Bath, UK, and Senior Research Consultant at Sangath Community NGO, Goa, India. Richard is both a clinical and an academic psychologist. He is a leading authority on substance misuse, with a special interest in the impact of this misuse on other family members, including children. He is a founder of the Alcohol, Drugs and the Family UK research network and AFINet (Addiction and the Family International Network: www.afi-network.info/) and member of the 15-person Scientific Committee of the EMCDDA (European Monitoring Centre on Drugs and Drug Addiction). He has published widely on substance use and substance use bereavement. He was a consultant for the ESRC-funded bereavement through substance use research on which this book is based.

Tony Walter is Honorary Professor of Death Studies at the University of Bath and has taught, researched and published on social aspects of death, dying and bereavement for 25 years. He was Director of the University of Bath's Centre for Death and Society from 2011 to 2015. Tony continues to work with CDAS following his retirement. He also works with the churches and Civil Ceremonies, Ltd. to train funeral celebrants. He has authored 11 books and edited 3 and has published over 100 academic articles and chapters. His books include *Funerals and How to Improve Them* (Hodder), *The Revival of Death* (Routledge) and *On Bereavement* (Open University Press). He was the Principal Investigator for the ESRC-funded research on which this book is based.

Researching families bereaved by alcohol or drugs

Christine Valentine and Linda Bauld

This book presents findings from a large qualitative study of bereavement following a family member or other close person's death as a result of substance use. Funded by the UK Economic and Social Research Council, the study was conducted by researchers based in the Centre for Death and Society (CDAS) at the University of Bath, in collaboration with the University of Stirling; the research was carried out in England and Scotland over three years, from 2012 to 2015. Although potentially devastating (Feigelman et al., 2013), substance use bereavement has hitherto received little attention in the academic literature. There has been no evidence-based guidance to help addiction or bereavement services to support people bereaved in this way.

The study was inspired by a bereaved mother, Joan Hollywood, whose son died through substance use and who campaigned to raise awareness of, and improve support for, this kind of bereavement. It was Joan who brought our attention to the lack of support and understanding for these bereaved people, and we were particularly fortunate that she agreed to be a member of the research team. Indeed, her involvement was crucial to the study both theoretically and practically. Joan died just a few months before the end of the study, and what follows is dedicated to her memory.

Although affecting a sizeable number of people worldwide, substance use bereavement has been internationally neglected in research, policy and practice. In this introduction we first consider why the topic and the research are important – because of how many people are affected by such deaths and because they are a particularly vulnerable, stigmatised and 'at risk' group (Valentine et al., 2016). Second, we provide a brief overview of our aims for the study, the collaborative approach we adopted, the methods we used – in-depth interviews with bereaved adults, focus groups and a working group of practitioners – and how these three methods built on and informed each other. We draw attention to how the research benefitted from:

*   involving a wide range of stakeholders, including bereaved individuals, practitioners and policymakers from a range of services in both the public

sector (health, social care and criminal justice) and the charitable sector (substance use and bereavement services);
- a collaborative approach to accessing people's experiences, which could inform and improve support for this group.

Why the research is important

Estimates of how many people are affected by substance use deaths

This book is based on the first study to examine, in depth, the experiences of 106 adults bereaved following a drug- and/or alcohol-related death. Estimates of the prevalence of these deaths worldwide suggest that the numbers of bereaved individuals and families left behind are considerable. The most recent estimate of alcohol-related deaths worldwide is 2.5 million per year (WHO, 2013), making alcohol the world's third-largest risk factor for premature mortality and morbidity. In 2010, deaths as a result of illicit drug use amounted to between 99,000 and 253,000 globally, or between 22 and 56 deaths per million of the population aged 15 to 64 (UNODC, 2012). In North America and Oceania drug-related deaths account for approximately 1 in every 20 deaths among persons aged 15 to 64; in Asia approximately 1 in 100 deaths; in Europe 1 in 110; in Africa 1 in 150; and in South America approximately 1 in every 200.

The numbers of drug-related deaths in both Britain and the United States are on the rise (ONS, 2015; National Records of Scotland, 2014; National Institute on Drug Abuse, 2015). Across Great Britain, with a population of 64.4 million (ONS, 2015), in 2014, 3,346 drug-related deaths were registered in England and Wales (ONS, 2015) and 613 in Scotland (National Records of Scotland, 2016), in both cases the highest ever recorded. In the same year, alcohol-related deaths were 8,697 (Office for National Statistics, 2016), with 7,545 of these being registered in England and Wales and 1,152 in Scotland (National Records of Scotland, 2013). Although alcohol-related deaths have fallen since 2008, they are still higher than in 1994 when comparable records began. However, both drug- and alcohol-related deaths are likely to be far more common than official statistics convey, because not all these deaths are recorded or categorised as being alcohol or drug related, and how each of them is defined tends to vary (Gossop et al., 2002; Beynon et al., 2010).

Despite these statistics, substance use bereavement has been internationally neglected in research, policy and practice. With policy and practice initiatives focused on preventing such deaths and treating the cause, those left behind may be considered part of the problem, an unwelcome reminder of the limits to prevention and treatment (Valentine and Bauld, 2016). However, their experiences also suggest that substance use may be a far more complex and nuanced issue than is apparent.

International evidence of the vulnerability of those bereaved by substance use

These statistics are worrying, not only in representing lives cut short, but also in reflecting the considerably greater number of those who have been left behind and, as evidenced by the small amount of previous research, represent a particularly vulnerable, stigmatised and 'at risk' group in terms of health and well-being (Valentine and Bauld, 2016). Our literature review (Valentine et al., 2016) identified just three small qualitative research studies and one survey. A study from Brazil interviewed six bereaved family members (Da Silva et al., 2007), and a pilot study from England interviewed four bereaved family members (Guy, 2004; Guy and Holloway, 2007) about the drug-related deaths of their children. More recently a third study interviewed four British teenage girls, aged between 14 and 16 years, about the deaths of their parents, including deaths that were alcohol related (Grace, 2012). These three studies found that grief may be compounded by the stresses and strains of having lived with the person's substance use and the stigma associated with these deaths (see Chapter 2). In addition, in the United States, Feigelman et al. (2012) undertook an exploratory, comparative survey of 571 bereaved parents whose children had died from different causes, 48 being drug related, 462 from suicide, 24 from natural causes and 37 being accidental. They concluded that, although both suicide and drug-related deaths were particularly difficult to grieve, parents bereaved by drug-related deaths suffered greater stigmatisation and associated lack of compassion than did those bereaved by suicide.

Study design

Our research design adopted a qualitative approach informed by the literature. Here we set out the study aims and the collaborative approach employed throughout the study.

Aims

To build on and extend existing research, our aim was two-fold: first, to gain both an in-depth and a broad understanding of what these bereaved people were coping with in order to raise awareness; second, on the basis of this understanding to consider how existing service provision might better meet the needs of the bereaved people concerned. To address these aims we adopted a three-phase model of research, using different methods for each phase.

Phase One involved mainly individual, face-to-face interviews with 106 adults bereaved by substance use (although six of the interviews were with married couples and seven interviews were conducted via telephone); we defined substance use broadly to capture any death where drugs or alcohol were believed to be involved, including suicide and accidents. Phases Two and

Three involved focus groups and a working group, respectively, mainly comprising practitioners whose work brought them into contact with substance use deaths and/or bereavement, although also including some people bereaved by substance use. This approach was iterative in that the findings from Phase One interviews informed the focus group discussions and the working group's task of developing best practice guidelines for service improvement. Phase Two focus group findings provided a further perspective on the experiences reported in the interviews, which in turn fed into the working group task.

Collaborative approach

Both the topic of the research and the aim to address a real-world social issue involved bringing together different groups of people, not only those participating in the study but also those who were conducting it. The topic provided the opportunity for collaboration between researchers from the two otherwise separate disciplinary fields of addictions research and bereavement. The study itself brought together bereaved adults directly affected by substance-related deaths; in the project, they collaborated in promoting improvements to the services the bereaved were likely to come into contact with.

The researchers formed a team of nine members combining the expertise of two death studies academics specialising in bereavement, six social scientists with expertise in addiction and, as indicated, Joan Hollywood, a bereaved family member and practitioner with experience of both fields. The two bereavement researchers and two of the addiction researchers were based at the University of Bath, three other addiction researchers at the University of Stirling and the remaining addictions researcher at the Liverpool John Moores' University. With two disciplines and three universities we had access to a wide range of mainly academic resources and networks. In addition, Joan's membership of the team provided access to networks that were to prove crucial in putting us in touch with both the bereaved people and the practitioners we needed to speak to. As the founder of a local support organisation, Bereavement Through Addiction (BTA),[1] Joan had forged links with others bereaved by substance use, as well as with a range of relevant local and national organisations.

This collaboration enabled us to make the study known to people we wished to interview, that is, adults bereaved by substance use, via drug and alcohol treatment services, generic bereavement services and the very few services offering specific support for this type of bereavement. It also enabled links with professionals and practitioners from a range of public, private and charitable sector services. Our initial method of approach was to organise and host an introductory event in Bristol, the benefits of which only became fully apparent at the end of the study. At the time, we hoped that this would be an effective way to make the study known and enlist the support of relevant service providers, as well as some service users in the southwest of England (we did not hold a similar event in Scotland where it felt more appropriate and cost effective to approach relevant contacts directly). Invitations to the Bristol event were sent to 60 people, informing them that we were keen to receive their input into designing the research. Fifty people attended the event, which included the following activities:

1 A presentation about how we proposed to carry out the research over the three years, including the aims and objectives and who would be involved, emphasising the important role that service providers could play. We also drew attention to the project having been inspired by a bereaved family member/service provider, who identified the lack of understanding and support for this type of loss.

2 A presentation by Joan, the bereaved family member on the research team, together with her husband, in which they shared their own experience of losing their son through substance use and how, as a result, they set up BTA, a charitable organisation in Bristol, providing support groups for family members and training/workshops for services who come into contact with this type of bereavement. The organisation has links with other organisations locally and nationally, provides a helpline and holds an annual memorial event in Bristol.

3 A group work session with attendees designed to elicit their experiences and views about the relevant issues involved and how to improve service responses to this kind of bereavement, as well as what they hoped would come out of the research. The session was guided by three main questions:

    i   What do you think are the key issues faced by family members affected by an alcohol- and/or drug-related death?
    ii   What do you think are the key issues which services need to consider in supporting family members affected by an alcohol- and or drug-related death?
    iii What are your experiences of contact with and supporting family members affected by an alcohol- and/or drug-related death?

    Each group was provided with flip chart paper to record and summarise its responses and agree on four key points to provide feedback to the larger group. We retained and subsequently collated, studied and used these responses to inform our interview schedules as well as our own understanding. In addition to the value of the information that emerged, the group activity enabled us to observe relationships between different services, between practitioners and bereaved people, and learn something of the dynamics involved in negotiating this complex and challenging area.

4 A session explaining our recruitment plans and our desire to work with a number of organisations that would be able to support the project and to assist with recruitment. This included raising awareness of the research

locally; arranging follow-up meetings and guidance to agree to a recruitment strategy that would be most suitable for that organisation and gaining individual organisational approval if needed; and helping with the dissemination of the findings, in particular the practice guidelines, which various organisations agreed to host on their websites.

With Joan's help, the attendee list for this event, to which we added throughout the study, provided an invaluable source of key contacts. We drew on these, not only for recruiting interviewees, but also for the focus and working groups, as well as for compiling a list of invitees to the event at which we launched the guidelines that emerged from the study (see Chapter 7). More specifically, the event benefitted the research by 'kick starting' the recruitment of bereaved people to interview. The first interviewees had attended the event, and further interviewees were referred by services represented at the event. The event also enabled us to access a further source of interview space. Based on previous experience we were anticipating that interviewees would not necessarily wish to be interviewed in their homes. As a result of our introductory event we were able to provide the further option of meeting in drug treatment/recovery centres as well as at the university. The centres had the advantage of being accessible, familiar and reassuring to some interviewees. The event also provided us with contacts for focus group recruitment, with these early contacts going on to introduce us to other services. By being eager to work with service providers, the research team was able to establish rapport with key service representatives early on in the project, crucial to Phases Two and Three.

Further elements of our collaborative approach included:

- The involvement of an advisory group of six practitioners working in the areas of bereavement or substance use, three from England and three from Scotland. Regular meetings with this group proved invaluable for ongoing direction and practical guidance.
- Displaying project information on the CDAS website and via the CDAS online newsletter (sent to 1,000 people globally) generated contact from members of the public, some of whom then elected to participate in the research as interviewees.
- A mutually supportive relationship with an existing Cruse/Adfam research partnership[2] (charities working with bereavement and families affected by addiction, respectively) enabled the two projects to complement and learn from each other.
- Project newsletters informing practitioners and those who took part in the study of the project's progress conveyed our appreciation of their interest and support. These included our developing findings from the interviews and focus groups, conference presentations, journal publications and the practice guidelines (see Chapter 7).

Methods

Interviews

To meet our aim of gaining both depth and breadth of understanding, we needed to access bereaved individuals who were prepared to talk to us at length about their experiences within their relevant national context (England and Scotland). Scotland's legal system differs from England's, including their systems of investigating sudden and/or unexplained deaths, which is the responsibility, respectively, of the Crown Office and Procurator Fiscal Service. Therefore some differences are apparent in interviewees' experiences of the statutory procedures in the immediate aftermath of the death.

We could not approach potential interviewees or target specific groups directly, because we wanted no one to feel pressured. This would have been both unethical and counter-productive, in view of the nature of the topic and the importance of engaging interviewees as fully as possible. We needed to rely initially on convenience sampling, or people volunteering and making themselves available to be interviewed, either initiating contact with us directly or with the help of a third party. However, once interviewing was underway, we were able to use some snowball and, later, purposive sampling to increase diversity; for example, we were able to interview bereaved individuals who were themselves in treatment for, or in recovery from, substance use. The final sample was diverse in age, relationship to the deceased, time since death, cause of death and personal experience of substance use (see Figure 1 and Chapter 5). However, it was not as ethnically diverse as we had hoped.

Before being interviewed, those who volunteered were given an information sheet and asked to sign a consent form. Interviewing took place between March and September 2012 and, apart from the seven telephone interviews, either in the researcher's or participant's work setting or in the participant's home. Interviews lasted between 40 minutes and more than two hours, and all were digitally recorded with interviewees' permission and then transcribed. Prior to the main interview, demographic data about both the interviewee and the deceased were collected.

The four interviewers (two in England, two in Scotland) adopted an open-ended, conversational approach, which both encouraged interviewees to tell their stories and focused on key areas, including the relationship with the deceased person before they died, the nature of their substance use, the circumstances and impact of the death, finding support and memory-making. Interviewing took account of the potential emotional distress of recalling painful experiences both during and following the interview, interviewers being sensitive to any fears participants had about being stigmatised, potentially discouraging them from sharing personal material. An open-ended approach to interviewing allowed participants to disclose only as much as they could manage and, as it turned out, almost all of them appreciated being able to share their stories. Indeed, other researchers have reported how bereaved interviewees found such sharing to provide relief and reinforcement (Parkes, 1995; Walter, 1996; Riches and Dawson, 1996; 2000; Valentine, 2007).

Data analysis was thematic, began during the latter stages of data collection and combined a grounded theory approach with interpretive phenomenological analysis so as to capture the participants' lived experience (Charmaz, 2003; Smith and Osborn, 2003). The data were imported into QSR Nvivo 10 and a coding framework developed using an iterative approach. Initially four members of the research team (the interviewers) independently read a sample of transcripts to identify key themes. These preliminary themes were subsequently discussed among the whole team and a draft coding framework developed and tested with a subset of 10 interviews. Further discussions around the suitability of the framework led to modifications and additions, leading to further testing until the final framework was agreed and applied to all the interview transcripts. Analysis was also supported by memos which summarised individual interviews or key themes.

Focus groups

With the benefit of Joan's contacts, we were able to identify potential focus group participants from a growing network of practitioners working for services from the public, private and charitable sectors who were showing interest in the research. These included individuals working for health and social care services, the police (including Police Scotland), the coroner's service, the procurator fiscal depute (Scotland), the funeral service, the press, the clergy, Public Health England, the Drugs Policy Unit, bereavement counselling, bereavement support and alcohol and drug treatment services. Those we identified were invited to attend one of four focus groups held in two locations in southwest England or one of two groups in Scotland, six groups in total. A total of 40 individuals agreed, each group having between five and eight representatives of different services – some of the practitioners had also been bereaved by substance use – and some bereaved people. Focus group discussions were informed by questions arising from the interview material and illustrated by interview quotes. These were sent out in advance to members, giving them time to reflect and prepare responses to the questions in light of their own experience. Mirroring findings from social work practice (Broadhurst, 2015), the use of interview quotes enabled us to bring bereaved people's experiences into closer view for practitioners and helped humanise discussion of practices and procedures. Because some practitioners' responses to those bereaved by substance use were at issue, the questions were designed to draw practitioners' perspectives and experiences.

In bringing together practitioners from a range of services, there was potential for both a clash of organisational cultures and mutual learning. Indeed, we discovered how services were fragmented and lacked knowledge of each other's work. Yet we were also struck by how much practitioners appreciated and took the opportunity for mutual learning; conversations continued and contact details were exchanged after formal discussions had ended. Otherwise, the focus groups both confirmed what we had learned from the interviews about how this group of bereaved people may be treated by UK services, while also providing insight into how professionals and practitioners working for those services struggle to work in an unwieldy and fragmented system (see Chapter 6). The focus groups proved particularly valuable in raising our awareness of practitioners' experiences, including the pressures and challenges of multi-agency working, as well as frustration and concern about how they were failing these bereaved service users. However, as we had hoped, they also took advantage of the focus group to share ideas about how services could improve.

The relevance of such a broad spectrum of services to the experiences of families in the immediate aftermath of a death, as indicated later and further discussed in Chapter 6, turned out to be both more crucial and more complex than we had anticipated. Indeed, the practitioners' side of the picture proved invaluable to the research through not only the focus group discussions, but also the working group and ultimately the best practice guidelines this group developed, as discussed in Chapter 7.

The working group

The inter-professional working group of 12 members included a paramedic, two members of Police Scotland, a senior coroner's officer, a general practitioner, a funeral director, a university chaplain, a senior alcohol policy and research officer, a counsellor and trainer in counselling and social care (who chaired the group) and three people who, in addition to working with substance users, were also bereaved by substance use. Most had taken part in the focus groups. This working group exemplified collaborative working (see Chapter 7), facilitated both by the chair's expertise in group process and the energy, commitment, cooperation and experience that group members brought to their task. This was to produce content for guidelines to assist professionals and practitioners from a range of services to respond better to those bereaved by substance use. The best practice guidelines (Cartwright, 2015), developed by the group and authored by the chair, present five key messages that emerged from both the interviews and the focus groups (see Chapter 7). Representing the combined expertise and experience of a range of practitioners whose work brought them into contact with this kind of bereavement and who, in some cases, were also bereaved themselves, the guidelines are both evidence and practice based. Developed by practitioners for practitioners, they have considerable potential to improve how those bereaved by substance use are treated by the services they encounter in the aftermath of such potentially devastating deaths. The guidelines are testimony to the willingness of those concerned to tackle this challenging and complex issue.

The structure of the book

The book has two parts. The larger Part I, of international significance, reports the experiences of adults bereaved by substance use as told to us in the interviews. The first four chapters are structured by four factors that make these deaths so difficult to grieve (see Introduction to Part I). Chapter 5 considers the diversity of people's experiences, comparing the experiences of different groups within the main sample, for example, different relationships with the deceased, differences related to the gender of the bereaved and different causes of death. By drawing attention to diversity, this chapter provides an important alternative perspective to the tendency to stigmatise and therefore homogenise this group of bereaved people. All these chapters are illustrated with interview extracts.[3]

Part II focuses on the role of services in dealing with substance-related deaths and is particularly relevant to UK policy and practice. It discusses the insights that emerged from the focus groups and the working group. Chapter 6 examines the findings from the focus groups. Chapter 7 discusses the five key messages we identified from both practitioners' and bereaved peoples' experiences and how these informed the best practice guidelines developed by the working group. Finally, Chapter 8 draws together the threads of the previous chapters to reflect on the study's main findings, the methods used and the key output (i.e., the practice guidelines).

Notes

1   BTA continues to provide a support group and helpline for people bereaved by substance use.
2   The BEAD (Bereaved Through Alcohol and Drugs) Project www.adfam.org.uk/professionals/latest_information_and_events/current_projects/bereavement
3   All chapters containing interview extracts adopt the following style: the relationship of the interviewee and the deceased (e.g., Mother/Father/Sister/Nephew, etc.) and whether interviewed in England (E) or Scotland (S) indicated in brackets at the end of each extract.

References

Beynon, C., McVeigh, J., Hurst, A., and Marr, A. (2010). Older and sicker: Changing mortality of drug users in treatment in the North West of England. *International Journal of Drug Policy*, 21, 429–431.

Broadhurst, K. (2015). Qualitative interview as special conversation (after removal). *Qualitative Social Work*, 14(3), 301–306.

Charmaz, K. (2003). Interpretative phenomenological analysis. In Smith, J. (Ed.). *Qualitative Psychology: A Practical Guide to Research Methods*. London: Sage, pp. 81–110.

Da Silva, E. A., Noto, A. R., and Formigoni, M. L. (2007). Death by drug overdose: Impact on families. *Journal of Psychoactive Drugs*, 39(3), 301–306.

Feigelman, W., Jordan, J. R., McIntosh, J. L., and Feigelman, B. (2012). *Devastating Losses: How Parents Cope With the Death of a Child to Suicide or Drugs*. New York: Springer.

Gossop, M., Stewart, D., Treacy, S., and Marsden, J. (2002). A prospective study of mortality among drug misusers during a 4-year period after seeking treatment. *Addiction*, 97, 39–47.

Grace, P. (2012). *On Track or Off the Rails? A Phenomenological Study of Children's Experiences of Dealing With Parental Bereavement Through Substance Misuse*. Unpublished PhD Thesis. University of Manchester, UK.

Guy, P. (2004) Bereavement through drug use: Messages from research. *Practice*, 16(1), 43–54.

Guy, P., and Holloway, M. (2007). Drug-related deaths and the 'Special Deaths' of late modernity. *Sociology*, 41(1), 83–96.

National Institute on Drug Abuse. (2015). Available at: www.drugabuse.gov/publications/drugfacts/nationwide-trends Accessed 13th April /2016.

National Records of Scotland. (2014). *Drug-Related Deaths in Scotland in 2013* [online]. Available at: www.nrscotland.gov.uk

ONS. (2016). *Alcohol-Related Deaths in the United Kingdom, Registered in 2014*. London: ONS. Available at: www.ons.gov.uk

ONS. (2015). *Deaths Related to Drug Poisoning in England and Wales: 2014 Registrations*. London: ONS. Available at: www.ons.gov.uk

Parkes, C. M. (1995). Guidelines for conducting ethical bereavement research. *Death Studies*, 19(2), 171–181. doi.org/10.1080/07481189508252723

Riches, G., and Dawson, P. (1996). Making stories and taking stories. *British Journal of Guidance and Counselling*, 24(3), 357–365.

Smith, J., and Osborn, M. (2003). Interpretive phenomenological analysis. In Smith, J. (Eds.). *Qualitative Psychology: A Practical Guide to Research Methods*. London: Sage, pp. 51–80.

UNODC. (2012). *World Drug Report*. United Nations publication, Sales No. E.12.XI.1.

Valentine, C. (2007). Methodological reflections: The role of the researcher in the construction of bereavement narratives. *Qualitative Social Work*, 6(2), 1–22.

Valentine, C., and Bauld, L. (2016). Marginalised deaths and social policy. In Foster, L. and Woodthorpe, K. (Eds.). *Death and Social Policy*. Basingstoke, New York: Palgrave Macmillan, Ch. 7, pp. 110–130.

Valentine, C., Bauld, L., and Walter, T. (2016). Bereavement following substance misuse: A disenfranchised grief. *Omega Journal of Death Studies*, 72(3), 283–301.

World Health Organisation. (2013). *Management of Substance Abuse*. Available at: www.who.int/substance_abuse/facts/psychoactives/en/index.html

Part I

Coping

Christine Valentine and Linda Bauld

The sample

To set the context for Part I, we describe the sample of bereaved people whose experiences form the basis of the following five chapters, illustrated by the table (see Figure 1). Representing 94 individuals and six couples, this is the largest known sample worldwide of this kind of bereaved persons; 66 interviews, which included the six couples, were conducted in England and 34 in Scotland. Most of those we interviewed were female, mainly mothers, particularly in Scotland. The deceased people they talked about were predominantly male, again particularly in Scotland, where they were mostly sons. Both bereaved and deceased were mostly White British. At the time of being interviewed, one-fifth of the interviewees were themselves in either treatment for or recovery from their own alcohol and/or drug use. When interviewed, the time since the death varied considerably from just a few weeks to over 30 years.

Causes of death

The cause of death and the role of alcohol and/or drugs in the death were wide ranging and hard to quantify for various reasons. In some cases the interviewee did not know exactly how the person died; in other cases there were discrepancies between the official cause of death and what the interviewee believed to be the cause of death. In one case, for example, the court delivered a verdict of drowning, whereas the mother believed it to be murder; in another case, the court delivered a narrative verdict because it was not clear whether the death was due to suicide or accident. In some cases alcohol and/or drugs were excluded from the official cause of death but believed by the interviewee to be implicated. Deaths involving alcohol and/or heroin or methadone predominated, and causes of death included overdose, suicide, illness (including substance-related conditions), accident, murder and manslaughter. However, the interview data suggest that around one-third of deaths resulted from a drug' overdose, usually involving heroin (see Templeton et al., 2016); roughly one-quarter of deaths could be directly attributed to alcohol; and approximately

	England (N = 66) Interviewees = 71 Deceased = 66	Scotland (N = 34) Interviewees = 35 Deceased = 36
Gender of interviewee	Female = 49 Male = 22	Female = 30 Male = 5
Mean age (years) of interviewee at interview	51 (range 22–75)	54 (range 23–75)
Gender of deceased	Male = 48 Female = 18	Male = 31 Female = 5
Mean age of deceased (years)	41 (range 16–84)	33 (range 16–80)

Relationship of
interviewee to
deceased
	Parent = 30
	Child = 19 (includes 1
	   adopted child)
	Spouses/partners = 9
	   (includes 1 ex-spouse/
	   partner and 1 LGB
	   partner
	Sibling = 9 (includes
	   2 step-siblings)
	Friend = 5
	Niece = 2
		Parent = 26
		Partner or
		   ex-partner/
		   spouse = 4
		Sibling = 4
		Child = 2
		Niece = 1
		Friend = 1

Figure 1 Profile of sample (N = 100 interviews)

15 deaths were classed as suicides. There were also two murders and one man-slaughter, a range of illnesses (including cancer, pneumonia, tuberculosis, hepa-titis C, food poisoning and sudden adult death syndrome), accidents (a fire, a road traffic accident, misadventure and drowning) and deaths from complica-tions related to drugs or drugs and alcohol use combined.

The findings

With the benefit of such a large data set, our interview findings break new ground in explaining the combination of factors that make these deaths so difficult to grieve and how these difficulties reflect socio–cultural norms about how death should be handled. Thus we learned that these deaths can be par-ticularly problematic for those left behind due to four main interactive factors, which can be summarised as 'the life', 'the death', 'the stigma' and 'the memory'. Each of these forms the basis of Chapters 1 to 4, as follows:

- Chapter 1 highlights the stresses and strains of having lived, in some cases for many years, with the substance use of someone close (see e.g. Copello et al., 2009; Orford et al., 2012).
- Chapter 2 discusses the difficult circumstances of the death, including how and where the person died and the often inadequate support and insensitive

treatment these bereaved people may receive from services mandated to deal with the death.

- Chapter 3 examines interviewees' experiences of and responses to the stigma of substance use (Walter et al., 2015), which may extend to the families and others associated with the person both before and after the death.
- Chapter 4 explores the difficulties of remembering and memorialising a life that may be considered by others (and, in some cases, the bereaved) as unfulfilled or even wasted.

Of these four factors, bereaved people's experiences of services (Chapter 2) were unexpectedly prominent, particularly in the immediate aftermath of the death when they were at their most vulnerable. These experiences drew our attention to the complex and fragmented multi-agency system within which these deaths are dealt and which these bereaved people – and, as we were to learn, the professionals and practitioners who were part of that system (discussed in Part II) – were up against. As a result this study is the first to consider how substance use bereavement is affected and often compounded by both those dealing with the death's immediate aftermath and those providing bereavement support.

These four factors raise issues of global relevance, but our account of how these deaths are dealt with by services is UK specific, although showing some differences between England and Scotland. However, whether global or nationally specific, and regardless of the four factors in common, our data found that individual experiences of this kind of death and loss, as with death and loss more generally, were nonetheless diverse, as discussed in the final chapter of Part I of the book. Chapter 5 identifies and discusses interviewees' diverse experiences of this kind of bereavement. It argues that recognising diversity is crucial in countering the stigma by which the individuals concerned are stereotyped and labelled as if they were all the same.

References

Copello, A., Templeton, L., and Powell, J. (2009). *Adult Family Members and Carers of Dependent Drug Users: Prevalence, Social Cost, Resource Savings and Treatment Responses. Final Report to the UK Drug Policy Commission.* London: UK DPC.

Orford, J., Velleman, R., Guillermina, N., Templeton, L., and Copello, A. (2012). Addiction in the family is a major but neglected contributor to the global burden of adult ill-health. *Social Science and Medicine*, 78, 70–77.

Templeton, L., Valentine, C., McKell, J., Ford, A., Velleman, R., Walter, T., Hay, G., Bauld, L., and Hollywood, J. (2016). Bereavement following a fatal overdose: The experiences of adults in England and Scotland. *Drugs: Education, Prevention and Policy*. Available at: www.tandfonline.com/doi/full/10.3109/09687637.2015.1127328

Walter, T., Ford, A., Valentine, C., Templeton, L., and Velleman, R. (2015). Everyday kindness: How people bereaved by alcohol or drugs experience compassion and stigma. *Health and Social Care in the Community*. Available at: http://dx.doi.org/10.1111/hsc.12273

Chapter 1

Families living with and bereaved by substance use

Lorna Templeton and Richard Velleman

Introduction

Having a relative or friend[1] who uses alcohol or drugs problematically is often extremely difficult and usually highly stressful. And it is not rare: it has been very conservatively estimated that globally 100 million adults are likely to be affected by their relatives' substance use problems (Orford et al., 2013). Considerable research has been undertaken, in a range of countries, on what this experience is like for affected family members (AFMs). A lot of this research has been undertaken by a research group based in the UK (AFINet-UK, formerly the Alcohol, Drugs and the Family Research Group, of which the authors of this chapter are members),[2] but many other researchers have also examined this issue (e.g., Barnard, 2007; Casswell et al., 2011; Philpott and Christie, 2008; Ray et al., 2009; Wiseman, 1991). Large numbers of detailed interviews have been conducted (in the research conducted by AFINet, more than 800 AFMs have been interviewed) and a considerable amount of quantitative questionnaire data has also been collected. Most of these research participants have been close relatives, and considerably more women than men have participated: for example, wives/female partners and mothers are the two groups most commonly represented in our research. But overall, the studies which have been completed have included a diverse range of relationships, including husbands/male partners, fathers, siblings, sons and daughters and sometimes extended family members like aunts, uncles and cousins.

The result of this research has been the development of a clear picture, relatively consistent across geography, socio-cultural groups and type of AFM, showing that AFMs experience multiple stresses, coping dilemmas and an overall lack of information and support. As a result, AFMs are at significantly heightened risk for ill health and other problems, which prove very costly, both personally and for public services (Copello et al., 2010c; Templeton, 2013; Ray et al., 2009).

Based on our many years of experience of undertaking research across the UK and a wide range of other countries and socio-cultural groups (e.g., Arcidiacono et al., 2009, 2010; Orford et al., 2005a, 2010a; Velleman and

Templeton, 2003), we and our colleagues have developed a model which summarises our understanding of how a close other's substance problems can affect AFMs so negatively. This is the Stress-Strain-Coping-Support (SSCS) model (Orford et al., 2010a, 2013), which, unlike many other approaches (described in Orford et al., 2010a), offers a non-pathological way of understanding their circumstances. The SSCS model, and the research underpinning it, suggests that:

1) Living in a family where someone misuses alcohol or drugs is commonly very **stressful**, both for the person misusing the substances and for anyone close to them. Substance misuse can and often does have a significantly negative effect on family life in general and on individual family members.

2) AFMs who are affected by and concerned about a drinking or drug problem in the family are likely to show signs of **strain**, including forms of physical and psychological ill health.

3) AFMs in this situation are often faced with a difficult life task in trying to understand what is going wrong and what to do about it (we refer to these ways of understanding and responding as '**coping**'); this can cause great dilemmas over what to do for the best.

4) A further issue facing AFMs is understanding both what is happening to their substance using relative and why it is happening. This understanding may also influence what sort of stance they feel they ought to take towards both the substance and the relative, which relates to the previous point about how they cope. Part of gaining a better **understanding** is receiving good, accurate **information**. Sometimes this is of a purely factual kind; for example, the names of types of illicit drugs, the means of their administration and some of their effects, or information about the strength of different alcoholic beverages. But many AFMs also often find it useful to discuss the nature of addiction or dependence and the difficulties their relatives have in overcoming it and how treatment works. AFMs may also require information and understanding about a range of other areas too, such as mental health problems, domestic abuse and social welfare or other financial issues.

5) AFMs can be helped or hindered in how well they respond to and how well they understand the problem by how other people react and interact. This is the '**support**' component, and the other people include other family members, friends, neighbours, colleagues and professionals.

6) The stress describes the impact of the problem drinking or drug use on the other members of the family, and this stress leads to strain. But for any given amount of stress, the amount of strain that is caused is influenced by the positive or negative effects of these three other factors: the information they receive, the method(s) of coping used and the level and quality of social support.

The general conclusion of our research has been that although there are differences, there is also a 'common core' to AFMs' experience[3] that consists of high levels of stress, a set of common coping dilemmas, difficulties in obtaining good quality social support and high levels of strain usually manifested through physical and psychological symptoms. We may term this as the 'burden' borne by AFMs (Orford et al., 2013). It has also been concluded that this common core pertains largely independently of the relative's addiction (alcohol or illicit drugs or gambling)[4] and independently of factors such as the affected family member's sex and relationship to the substance-misusing relative.

Having stated that this common core is largely independent of a range of factors, this must not be over-stated. The reality seems more to be that there *is* a core experience but also that some differences, albeit more minor ones, do emerge (e.g. Orford et al., 2016, 2005). These include how socio-cultural factors influence coping; gender roles in different countries and cultures, which affect the ways that AFMs both experience stress and demonstrate strain; and whether or not AFMs express (or even recognise) feelings of resentment and anger at how the drinking or drug problem has restricted family members' lives. Furthermore, the predominance of women in our research means that the universality with respect to men is not so clear, and the predominance of partners (usually female) and parents means that universality with respect to other forms of close family (and friend) relationships is also not so clear.

The focus of our study on bereavement through substance use provided an interesting opportunity for us to extend our programme of research by reflecting on the possible application of the SSCS model to bereavement. The first part of our semi-structured interview asked interviewees about their relative's or friend's substance use and the impact that this had on them (the interviewee), on others in the family and on the relationship between interviewees, wider families and the deceased. These often very lengthy accounts, along with the other areas covered in the interviews, provided a valuable opportunity to better understand the experiences of our bereaved interviewees before death and the impact which those experiences often continued to have after the death. In the following sections of this chapter we will discuss how the SSCS model can be applied to AFMs' experiences both before and after death.

Findings – applying the SSCS model before death

Analysis of our data shows how the SSCS model can be applied to our interviewees' accounts of living with their relative or friend's substance use,[5] in the majority of cases for considerable periods of time, before that person died. Reflecting the SSCS model, our interviewees described the stresses which they were often under and the resulting strain this brought for themselves and their families.

> *Fights were on a daily basis in the house when I was going through school . . . and he would always be shouting at my mum. . . . You could see it gradually falling apart, getting worse and worse as the years went on . . . it was horrible to see that happen to someone you love. . . . It obviously did affect me . . . I remember finding school very*

difficult, finding socialising with people very difficult . . . [I] didn't have a normal life like most other people did and I had panic attacks and things like that . . . I felt quite scared of him really, quite intimidated . . . Me and my mum . . . we were just desperate really. We'd cry together sometimes. . . . We were depressed really . . . because it just used to be the same thing every day.

(SisterE)

Some interviewees also explained that they found it hard to understand why their loved one was using alcohol or drugs so destructively.

But to actually be told that he had a drug addiction I was like how come, he is working . . . how can you hold up a full time job and be a drug addict. It didn't make sense to me . . . maybe I was a bit naïve because I had never really had a lot of experience of [that] . . . I didn't really understand the ins and outs of a drug addict and what it involved and I wasn't as knowledgeable as probably I could have been.

(FriendS)

Following on from this, interviewees also explained the difficulties they faced in working out how to cope with the stresses and strains that they faced.

I took to treading [on] eggshells, pussyfooting around [my son], so then we weren't really having a real relationship, because I knew it was going on and he knew that I knew it was going on, but I wasn't prepared to tackle him about it. So I took the coward's way out, so [my son] would go to his room. I knew he was taking a drink up there with him and whatever else he might be doing. And then I just let him stay in his room and then I would just check on him every now and again, make sure he was eating, make sure he was alive. I didn't know what to do . . . And what we didn't know . . . he was moving into harder drugs . . . and we kept him short of money.

(MotherE)

It was because of his drinking that I started to challenge him on it and then we had a very, very, very difficult relationship, because I don't know if the two just coincided as a coincidence, but his drinking then started getting worse, which then made me even more – I think I was just angry at him, like why – you know, who does he think he is that he can just do this to a family?

(DaughterE)

Finally, it was clear that a significant number of our interviewees lacked formal and informal support in dealing with the problems that they faced, with some able to reflect on how hard it can be to seek help.

I didn't know there was such a thing until I went in and found out there was family support . . . we have had the drug worker in the house working with [our son] and he never even mentioned family support.

(FatherS)

At that stage it was all out in the open, but his family refused to believe that he had a problem, absolutely refused to believe it. So I found that very difficult doing it all on my own . . . I didn't have a soul, I've got some very good friends who I'd known for years and one in particular who I've known for 45 years and she said, "Why didn't you tell us?"

(WifeE)

I would say there probably wasn't a lot of support around but I do think that I and an awful lot of people in the situation that I was in isolate ourselves. There is a lot of shame in it and guilt I think and that feeling of it's all my fault kind of thing . . . I don't actually know how much help I could have accessed even if it had been there at the time.

(WifeS)

Overall, much of what our interviewees said mirrors the findings from our broader programme of research (described earlier) and from the wider UK and international literature (e.g., Bortolon et al., 2016; Esser et al., 2016; Fereidouni et al., 2014; Orford et al., 2005) about what it is like to live with someone who uses substances problematically or harmfully. However, although our interviewee accounts of how they were affected by the pre-death substance use were largely the 'back story' to our primary focus on their experiences and needs following the death, their stories offer useful additional insights into the stress and coping aspects of our model. These are things which have not previously emerged as dominant in our other research, mainly because the majority of participants in our previous studies had not been affected by the death of their loved one. Figure 1.1 depicts the original SSCS model (Orford et al., 2013) with our suggested additions (which appear in bold text or shaded ovals) which we summarise next.

Stress

Our data suggest that there are potentially three additions to make to this aspect of our model.

First, a sizeable number of the sample talked about their experience of previous overdose or suicide attempts (related to the person who later died), with some directly involved in responding to these incidents (e.g., resuscitating the person or calling the emergency services), demonstrating how close they had come to death before it actually occurred (see also Templeton et al., 2016b).

He had a history of overdosing, getting clean, overdosing, getting clean.

(MotherS)

Mum opened the back door and she heard the engine running in the garage, and he [Dad, an 'alcoholic'] had tried to commit suicide.

(DaughterE)

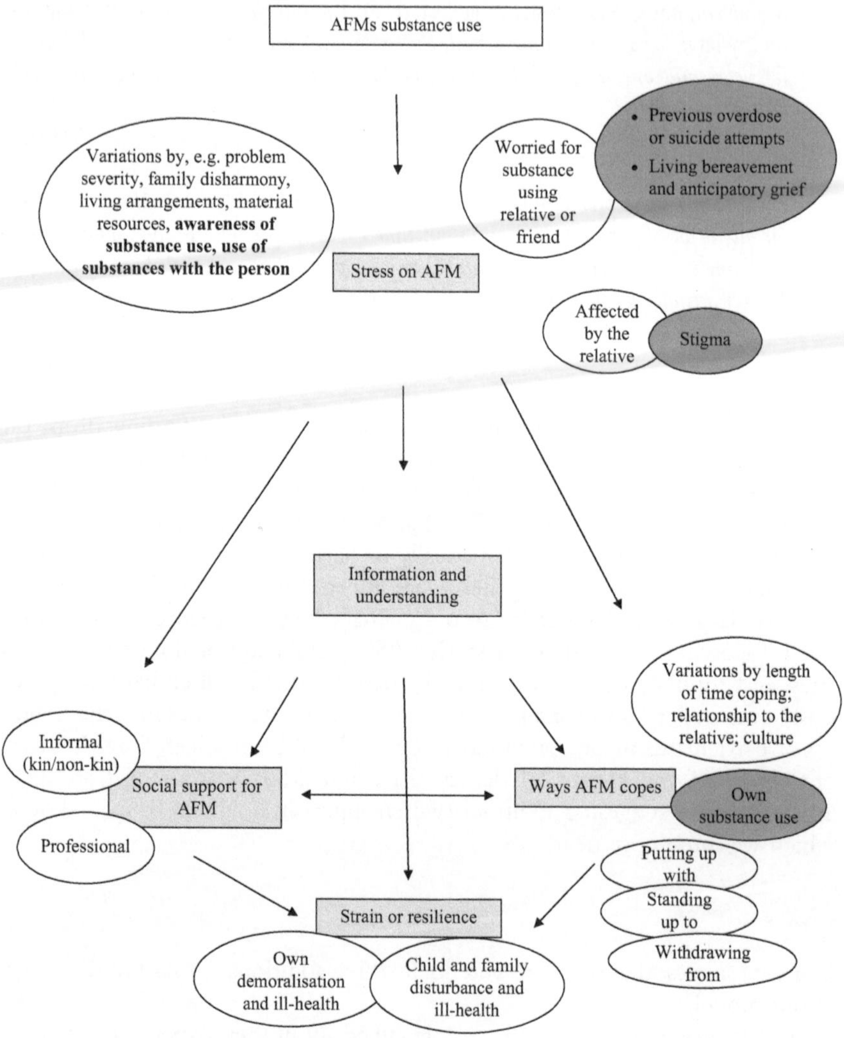

AFMs substance use

Variations by, e.g. problem severity, family disharmony, living arrangements, material resources, **awareness of substance use, use of substances with the person**

Worried for substance using relative or friend

- Previous overdose or suicide attempts
- Living bereavement and anticipatory grief

Stress on AFM

Affected by the relative

Stigma

Information and understanding

Informal (kin/non-kin)

Social support for AFM

Ways AFM copes

Own substance use

Variations by length of time coping; relationship to the relative; culture

Professional

Strain or resilience

Putting up with

Standing up to

Withdrawing from

Own demoralisation and ill-health

Child and family disturbance and ill-health

Figure 1.1 **Stress-Strain-Coping-Support model with consideration of adults bereaved by substance use and the impact before death (new additions in bold text or shaded boxes)**

Overlapping with this, many interviewees, when telling us about the impact of their loved one's substance use on them, suggested that they had already 'lost' the person that they knew before they died, what one mother called a "*living bereavement*". Some were also anticipating the person's death and the grief that this would bring for them (see also, for example, Da Silva et al., 2007; Templeton et al., 2016a, 2016b).

I was sad. . . [but] . . . I don't think I shed any tears for him though. I think I had done all that before.

(Ex-wifeS)

I really knew six months before she died she'd totally given up, that she'd accepted her fate.

(HusbandE)

I lost my mum when this (i.e., alcohol use) started. I always hoped I would have my mum back. So I grieved the loss of my mum [and then] I have a second grief for the person she became with her addiction.

(DaughterE)

Second, in the research discussed in this book, stigma emerges as important, although different categories of relatives appear to experience different levels of stigma (see Chapter 3). Many AFMs felt stigmatised 'by association' as family members of someone who misused substances while their relative was alive, and many also felt this stigma after the death in how both the deceased person and they as AFMs were treated by a range of authorities and others. Stigma has not hitherto been a central part of the SSCS model. This may be because when it arose as an issue in interviews or discussions, it was simply considered to be one of many variations of 'stress', and that is where we have placed it in our revised model. Indeed, whether or not to place stigma more centrally is still under discussion – it is not clear that the experience of stigma in the ways that an AFM is treated is a more harmful experience than (say) domestic violence from someone within the close family. On the other hand, it may have emerged as more central in the interviews for this present project because it was more openly and specifically asked about as part of experiences after death. Overall, it is clear that we need to carefully consider how to best include stigma in our SSCS model and to further explore the impact that it may have on AFMs' experiences and on the other components of our model such as support and coping.

Overall, our interviewees talked less about their experiences of stigma before the person's death, probably because the focus of the interviews was the death and experiences thereafter. Nevertheless, when it was discussed, what interviewees had to say mirrors other research which has investigated stigma in substance users and their families (e.g., Adfam, 2012; Lloyd, 2010).

When he was alive I did keep a lot of it a secret from my friends.

(SisterE)

Some of our neighbours, I'm sure they used to look at us and think, oh, that family, because there was lots of yelling and screaming at [our son] as he's going out of the door and he's telling us to eff off and all sorts of things. So we always had the police at our door. So you just create your own little opinion about somebody in their family, but you don't know really what goes on.

(MotherE)

I was embarrassed when I found out she was a drug addict because I thought it was only me in this world who had a drug addict for a daughter.

(MotherS)

Third, the stressors which AFMS are exposed to may vary in other ways. One is the extent to which AFMs are aware of the person's substance use, an issue highlighted in one of the few other studies which has been conducted in the area of bereavement through substance use (Da Silva et al., 2007). In our study interviewees were usually aware of the person's substance use, but the levels of stress varied. For example, many of those affected by alcohol use had high levels of awareness of the extent of the problem, having been exposed to the drinking and its effects, often for many years.

I've never known him to be sober, apart from the short stints in rehab. He was always drinking.

(NieceE)

I would have been probably four or five when she really started to use alcohol regularly, because I remember her being drunk at a time when I didn't understand that's what the problem was . . . it was while I was at primary school that she started drinking much more heavily habitually during the day. And certainly by the time I was in junior school . . . I was coming home [and] I could tell just by looking at her that she was drunk . . . as time went on and I went up to secondary school, it gradually got worse.

(DaughterE)

On the other hand, some of those exposed to drug problems had lower levels of awareness of the problem, either by not knowing for some time that the person was using drugs or being unaware that their death had been preceded by relapse.

She started taking drugs when she was eleven and I never knew because I was one of these mothers that said "Oh no my daughter won't take drugs". When I realised, I ate my words.

(MotherS)

Not only did we find [our son] dead, we found out as well that he was back on heroin.

(FatherS)

Another factor influencing AFMs' stress levels was whether or not the AFM was also using alcohol and/or drugs problematically themselves, either independently of or with the person who died. It is perhaps unsurprising that those who were themselves alcohol or drug users, and who might have used alongside the person who later died, experienced lower levels of stress about their relative's or friend's substance use than those where this was not the case. Nevertheless, this had not featured in our previous research samples, either because family members' substance use did not feature prominently in these studies or it was an exclusion criterion for our intervention evaluation studies. We therefore think it is worth considering stigma as an additional feature of our model. However, we did not inquire further about the impact of the other person's substance use on them when they were themselves a user. We cannot therefore say definitively that stress was absent for them, only that it appears to be a less important feature of their experiences when compared with AFMs who do not themselves have problems with alcohol or drugs.

Influence of diversity

We indicated earlier in this chapter that there are a number of dimensions along which the experiences of AFMs may vary, although little global research has examined these differences in depth. Our study offers the opportunity to consider variations for AFMs along some of these dimensions. Given the focus of our study and the scope of this chapter, we cannot examine this in detail so what we present are some ideas on what some of these variations might be for two of these dimensions, namely how the bereaved adult was related to the deceased, and whether the bereaved interviewee was themselves in treatment or recovery because of an alcohol or drug problem.

Although our previous research has included participants who represent a wide range of relationships with the problem substance user, parents, spouses/partners and now-adult children have predominated. The present sample has a greater representation of a range of other relationships which can be examined: we interviewed 12 siblings, six friends and three nieces, and so it is on these groups that we will focus here. Although it is unknown whether or not these experiences are representative, they are nonetheless an important starting point given that these groups have rarely been included in such research.

Although there was some variation in how strongly affected the 12 siblings were, many of them talked about quite major stresses and effects on them and on others in the family.

> *I began to notice that he was obviously using again . . . I was getting calls from phone boxes in those days to say that he'd overdosed in a phone box somewhere and I had no idea . . . he was just an absolute nightmare. It had a really big impact on the whole of the family for a long time . . . twenty odd years really. So yes, we had years of his drug abuse and him stealing off of us and the things that addicts do.*
>
> (SisterE)

There were times where he stayed with us and we would have the Police at the door . . . they arrive at your home at five in the morning saying we've got your brother locked up. That caused problems between my husband and I . . . I actually hadn't seen [my brother] for quite a while before he died. He had been in touch to get money . . . he said he had needed to leave [town] that it was trouble, there was people out to get him. And I had lent money and he hadn't left [town] and I suspect he had spent it on drugs.

(SisterS)

Many of the siblings also reinforced the themes of living bereavement and anticipatory grief discussed earlier.

I always knew that she was going to come to an end just because of my history with her . . . I had always been waiting, I'd been waiting for years for the phone call.

(BrotherE)

I was always convinced he would either completely end up with a complete irretrievable breakdown or that he would end up dead.

(SisterS)

I just felt like I lost my brother really.

(SisterE)

Although the three nieces all stated that the impact on them was less intense, in fact their descriptions show that the substance use of their uncle or aunt did have a significant impact.

Stress. . . . worrying about what was going to happen next . . . helpless . . . anger . . . guilt

(NieceS)

When he relapsed you would feel really disappointed in him. But because I was so young as well, I would get really angry at him for it . . . I understood he had a difficult life. It's not . . . you know, addiction isn't something that you can just get over. And I think that's what my family found hard to deal with is that it's no easy fix.

(NieceE)

Compared with the other sub-groups in our sample, the six friends (five males and one female) said very little about the impact of their friend's substance use on them. Possible reasons for this include the fact that four of them had a history of taking (and in two cases also dealing) drugs with their friend and described the strong relationship that existed between the drug use and their bond as friends, whereas another (the female friend) knew almost nothing about the drug use until a few months before death, and in the sixth case the friend died suddenly as a result of an alcohol-related accident when both the friend and interviewee were in their teens. Nevertheless, this group gave some insights into some of the struggles which they faced as a result of their friend's substance use. Two male friends were trying to come off drugs when the death occurred, and a third male friend expressed frustration at his friend for not being in recovery like he had been for the last few years.

> *I was angry with him because he didn't get into recovery. I felt disloyal . . . and I couldn't help him.*
>
> (FriendE)

The female friend talked of how she had struggled to understand her friend's addiction and him not conforming to her stereotypes of people with alcohol or drug problems.

Finally, when considering the 21 participants who were in treatment or recovery from their own substance use when they were interviewed, there was variation in how much this group talked about the impact of their relative's or friend's substance misuse and the nature of the stresses that they experienced. These narratives were particularly influenced by whether or not the interviewee had themselves used substances with the person who died and/or whether they were using when the death occurred. For those who were not in either of those categories, the nature of how they were affected mirrors that of the wider sample and the literature.

> *All of a sudden addict behaviour was there . . . if I'm truly honest I chose to ignore it . . . things weren't making sense, there were a few lies and this, and that and the other, so it was all those sorts of behaviours and again it caused friction in the relationship. I found that I was getting very stressed when normally I'm not a stress person really, things were getting on top of me . . . every night, I knew she was out drinking, is this going to be the night I am going to have a phone call.*
>
> (PartnerE)

Those who had used substances with the deceased still alluded to a range of possible stressors related to the other person's substance use, such as mental health problems, violence, prison and removal of children to the care system (all types of stresses which AFMs often mention as being associated with substance use). However, this group of bereaved people talked less about the impact which these experiences had on them than did other bereaved family members who had not used substances with the deceased. Further, some in this group through themselves being in treatment or recovery became more aware of the impact that substance use can have on others.

> *I've been on both sides now and I know the difficult side is being with somebody who drinks, not being the drinker, the drinking side is the easy side because you believe you are not doing anything wrong.*
>
> (Ex-wifeS)

I think that's why I cut down and got on top of myself because I didn't want my dad to go through the grief of having to worry about me drinking as well.

(NieceE)

In summary, a preliminary investigation of how our interviewees talked about the impact on them of their relative's or friend's substance use before that person died supports the application of our SSCS theoretical model to AFMs who are subsequently bereaved; some modifications to the model can incorporate these additional aspects to their experiences. We will now move on to applying the model to bereavement through substance use.

Findings – applying the SSCS model after death

Our study has provided a valuable opportunity for us to start to understand how the SSCS model might apply to what interviewees said about how their family member's substance use and subsequent death continued to affect them. Figure 1.2 shows a second version of the model, and next we unpack what some of the components of each aspect of the model might be when focusing on the specific experience of bereavement through substance use.

Stress

Our data suggest three broad ways that the death affects the stress, and hence grief, experienced by AFMs. First, through a range of circumstances present beforehand, death can continue to have a negative impact after death. These include the nature and impact of the substance use, the relationship between the bereaved and the deceased (see later and Chapter 5) and the bereaved's own problem substance use. One daughter explained the close association between experiences before death and the death itself, saying, "You can't just look at the death, because there's so much more build up to it".

In terms of the substance use itself, influencing factors include the duration and severity of the problem, the substance(s) being used (with perhaps greater stress attached to intravenous drug use and/or polydrug use) and whether other problems are also present such as mental health difficulties, unemployment or criminal activity. Any of these on their own or in combination can increase stress for the AFM after death, for example, because of the need to adjust to life after years of being exposed to substance use (and any additional problems), or because the stigma attached to substance use may shape conversations with other people about the death, among other aspects of grieving.

Grief could also be influenced by the relationship between our bereaved interviewees and their relative or friend while they were alive. Some spoke of close relationships and of how those relationships remained close despite substance-related stresses.

Our relationship is that she was my best friend . . . no-one knew her like me, certainly no-one knew me like her.

(BrotherE)

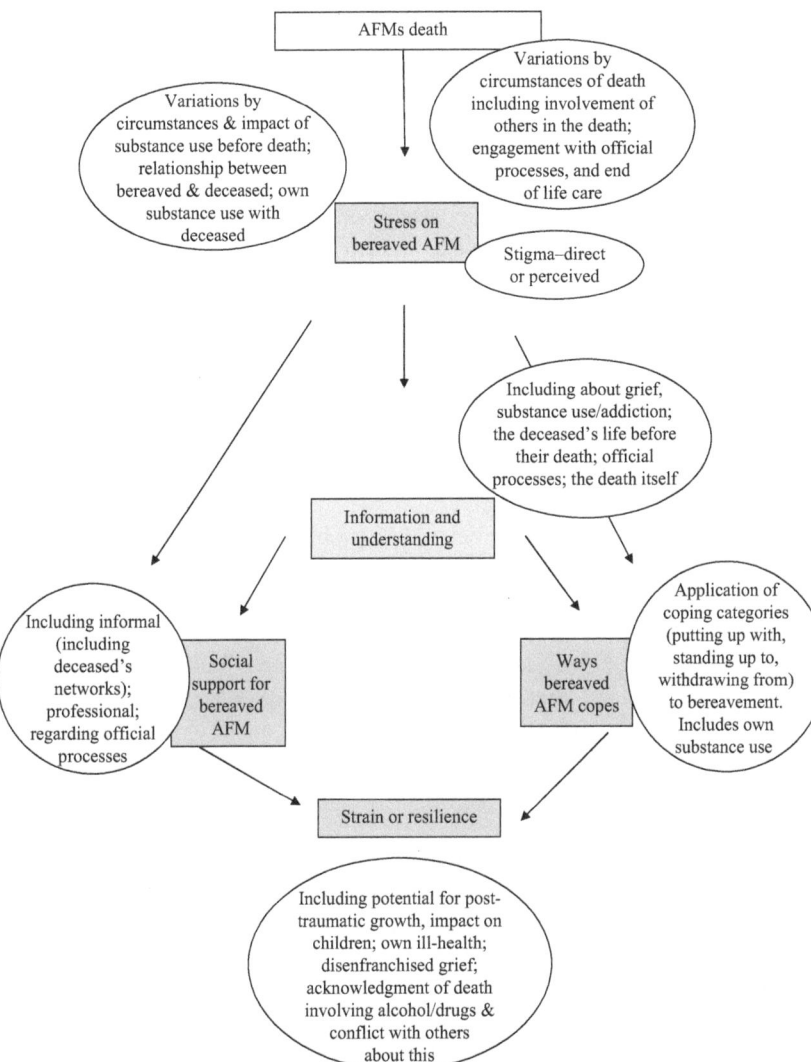

Figure 1.2 Stress-Strain-Coping-Support model with consideration of adults bereaved by substance use and the impact after death

I was really, really close to [my uncle] . . . Especially as I hadn't had the best child-hood myself, so he seemed to be the one that understood me the most . . . I lived about five minutes' away from him . . . I could pop in whenever I wanted. He could pop round whenever he wanted to. It was always really close.

(NieceE)

Others talked about very difficult long-standing relationships or relationships which became fractured as a result of the substance use.

> *I was really, really close to my dad . . . we used to do everything together . . . [but] . . . at one point we did cut him out of our lives because he was just too much to cope with . . . I kept my distance from him and eventually I slowly let him back in, but I had to keep him at arm's length . . . my relationship with my dad [was] very changed . . . [then] we started to rebuild our relationship with him, and unfortunately it was around about then that he started to get really ill.*
>
> (DaughterE)

> *There wasn't really much of a relationship to be honest . . . it was quite turbulent, because of his drug use. He used to steal off me . . . quite expensive things from my room . . . but also his moods were all over the place because of the drugs.*
>
> (SisterE)

Finally, some of those who themselves had problems with substances described continued or escalated levels of use after the death, which, for many, affected their experience of occasions like the funeral and affected their grieving.

> *[I] just used drugs to block everything out . . . and never really talked about it . . . I took drugs before I went [to view the body] . . . I got upset a bit [at the funeral], but because I took loads of drugs it just blocked it out for me so I was just able to get through it, put a front on and make out to my mum I was okay.*
>
> (BrotherE)

> *My problem was that on the day of the funeral . . . I just lost it, really. I thought I just can't get through the day. So I went to the fruit market, bought the stuff for the shop, and I stopped at a friend's shop on the way back and got him to give me half a bottle of gin. And I thought, well, I'll just drink this just to get me through the day. And then after that I'll stop. And I didn't. So I carried on drinking with the view [that] . . . if your dad had just died, you would be drinking. And it took me nearly two years to stop drinking. And it was probably the worst two years of my life.*
>
> (SonE)

Second, stress could also be influenced by a number of factors associated with the death itself. This includes how the death occurred, including end of life care, whether others were implicated or believed to be implicated in some way – this was most common with fatal overdose (Templeton et al., 2016b) and is explored elsewhere – and the extent to which the death required involvement of the authorities and official processes. In one case the interviewee believed their relative had been murdered through a contract killing and fought for some years to get a proper inquest.

The manner of the death could exacerbate stress in several ways, including knowing that the person died alone or in distress, finding out that the person was not found for some time (up to a few weeks in some cases), watching the person deteriorate and die in hospital or knowing that the person was murdered or had committed suicide.

> *The neighbours alerted that nobody had seen her for a few days . . . so the police broke in and found her . . . I still feel really bad that she died on her own.*
>
> (DaughterE)

> *And then over the seven days you saw him degrade literally. It was just taking over his body to the point where the last time I saw him his breathing was so far in between it was really, really scary. Like literally I thought at any second he was going to die. He was very, very yellow. He had a colostomy bag as well.*
>
> (NieceE)

Related to the manner of death was the presence or absence of good-quality end of life care and whether or not the relative or friend had died with dignity and at peace.

> *They got him breathing and took him in to the intensive care unit, and he was on a life support machine . . . eventually the consultant said we need to turn the machine off . . . So we agreed to having the machine turned off. And he died I think it was a day later . . . it was absolutely shocking – absolutely awful . . . he didn't slip away peacefully.*
>
> (MotherE)

> *I just thought this is so undignified . . . I definitely remember thinking that she was denied that because she was an alcoholic . . . regardless of anything, everybody deserves the right to die with dignity and I don't feel that my mum was given that.*
>
> (DaughterE)

We have reported elsewhere (Templeton et al., 2016a) that a sizeable number of the deaths required police involvement, a post-mortem and inquiries by the coroner (England) or procurator fiscal (Scotland). There were several ways in which these processes could add to or reduce our interviewees' stress, including how officials interacted with them (to be discussed later, under "Support"), delays with post-mortems and inquests and the impact of this on, for example, viewing the body, arranging the funeral or managing grief.

> *They kept her body for eight weeks . . . and we couldn't see her because of decomposing circumstances . . . I phoned them very day for eight weeks and they said "No, I am sorry".*
>
> (MotherS)

And that was painful, it took eight weeks before they released the body because it was deemed a suspicious death at first.

(PartnerE)

I think it was over a year [before the inquest] . . . I think I was quite surprised how much I was fretting . . . I think the suicide was bad but the fact you can't then move on until the inquest . . . I lost more weight coming up to the inquest then I did immediately after [my ex-husband] died . . . I wanted it to happen, I wanted some answers, I wanted resolution.

(WifeE)

Third, our analysis has highlighted that stigma, what one mother called a "*contaminated legacy*", is central to our interviewees' experiences and the associated stresses they faced. Stigma has been explored in detail elsewhere (Chapter 3; Walter et al., 2015) but in summary, it covers stigmatisation – both direct and perceived by the organisations and people our interviewees' came into contact with.

I was thinking "What are everybody in my work going to think of me? What are they going to think about my family? And what are they going to think of him?" . . . I was worried, what are they going to think of him as a person, thinking that he'd died of an overdose . . . it was in an area where we know a lot of people, so again that stigma issue came up for me, because I was worried about what people would think, I felt I had to justify his death to people; I had to explain what had happened and that, you know "he wasn't a drug user" and all the things that came with that.

(BrotherS)

Stigma is a significant part of understanding the experiences of adults bereaved by substance use and needs to be included in the adapted version of our theoretical model.

Strain

Our data indicate how the strain, which can manifest itself through a range of physical and psychological health problems, which AFMs may feel following the death of their relative or friend, can be exacerbated or lessened. This includes whether or not interviewees perceived their grief to be disenfranchised (Doka, 2002), how far the involvement of alcohol/drugs in the death was acknowledged and whether this caused conflict with others and the impact of the death on others including younger children.

First, particularly when interviewees experienced or perceived stigma towards themselves and/or their loved one, some felt they could not grieve openly or that the complicated emotions which they often experienced did not conform to what was expected of them.

Sometimes I feel like I'm trying to protect my friends, I feel like it's something people don't want to hear about or they're going to think bad of me or they're going to think oh your brother was a junkie – and I don't want to feel ashamed, I want to be able to just talk about it, but I think I do feel a sense of shame.

(SisterE)

I think the biggest thing with the way [our son] died was we didn't really feel entitled to grieve. Nobody said I couldn't, but because of the way he died, I felt that people felt, well, he was doing drugs. What are they upset about? And nobody ever suggested this for a minute. It was my perception.

(ParentsE)

Second, there was variation in how far interviewees felt able to acknowledge to themselves and others, both inside and outside the family, that the death was associated with alcohol or drugs. Direct or perceived stigma, or differences of opinion about how open to be about the death, could influence this.

To start off with I was embarrassed. The way he died . . . now, I am not, and that's because it's more the fact that he's died than how he died.

(MotherS)

I don't think anybody wanted to stand up and say anything. Because of the awkwardness and the circumstances . . . I don't think people wanted to address the situation even on his funeral day.

(DaughterE)

My aunt didn't want to tell anyone how my mum died, she wanted to say that she'd had a heart attack, she's so ashamed. I said why are we ashamed of this?

(DaughterE)

Third, bereaved AFMs could have heightened levels of strain if they had to deal with the impact of the death on others. For example, interviewees talked about a range of ways in which children, usually the deceased's siblings or children, were affected. This included their own grief, their lack of understanding about the death, including feeling that they were somehow to blame, and the cumulative impact of having been affected by the substance use before the death coupled with the effect of the death itself.

I still think my youngest son might be a sort of time bomb in a way because he's never really grieved at all.

(Ex-wifeE)

And (youngest son) had seen a needle and he kicked it under the drawers in the bedroom . . . and he thought that because he's kicked that under there (older son) *had got it again and that's how he died . . . he thought it was **that** needle . . . He thought he was to blame for killing [his brother].*

(MotherS)

And that has been really difficult that the two eldest haven't wanted to remember him and neither have I . . . I don't remember his birthday and there's no anniversaries. (Youngest daughter) has wanted to keep anniversaries and things, and I found that really difficult . . . Just (youngest daughter) finds it difficult to find ways to remember him.

(PartnerE)

Information and understanding

Interviewee narratives suggest several ways that the availability or absence of information and understanding in a range of areas can affect bereavement and grief. These include the relative's or friend's substance use and the nature of 'addiction' generally; retrospectively finding out more about the person's life before they died and gaining a better understanding of substance use to help come to terms with the death; the official processes with which many were involved; and wanting to have a clear picture about the death, such as its cause, whether others were present or involved and whether the person died peacefully or not.

I suppose when your mum on her death certificate has an alcohol related death it confirms to you that you really did have an alcoholic mother and so I guess it enabled me to feel I had the right to read books on alcoholic families, search the internet . . . for me I wanted to understand . . . why she drank I think. And I don't have the answer to that but I think [it's] helped [me] to understand a bit more about my family dynamic and why things are as they are.

(DaughterE)

The Coroner started summing up and I [wanted to ask a question but] . . . he said that is outside the remit of this Court . . . I thought inquest means having the answers . . . I mean you might have more realistic expectations, but it's partly because I didn't get given any information.

(WifeE)

I said to the [pathologist] he didn't suffer did he? I wouldn't like to think he was lying there gasping for breath and he was paralysed and he couldn't move. She said no . . . he would have slipped away.

(MotherS)

Coping

Our research about how AFMs cope with the substance use of a significant other has suggested three broad ways in which they endeavour to respond, namely putting up with it, standing up to it and withdrawing from it (Orford et al., 2010**a**, 2013). However, AFMs never fit neatly into any one box but fluctuate between styles of coping depending on the circumstances and resources available to them. Moreover, some AFMs may interpret a coping response in one way (putting up with it) but others interpret the same response another way (withdrawing). In considering what coping might mean after death, we need to think differently because many of the coping responses in our research involve some kind of interaction between an AFM and the person using substances, but this will obviously no longer apply after the death. We therefore now consider whether the three broad categories of coping can apply to the experiences of AFMs after death.

Our first broad category of coping, *putting up with it*, entails AFMs tolerating what they were dealing with, often out of fear that disrupting or changing the status quo might be even more stressful, or sometimes out of fear of what would happen if they attempted to respond differently or more assertively. When bereaved through substance use, *putting up with it* could cover feeling powerless to grieve or going along with others who want to hide the truth of the death or not talk about it (although for some this could be seen as a form of coping by withdrawal – see later).

> *You feel reticent to say my son died of a drug overdose because it tells people so much. They think it's telling them everything about that person and it's not.*
>
> (ParentsE)

> *When [my brother] was actually using and doing these things, [my mother] would talk about it to a certain extent, but since his death she doesn't, it's as if it never happened and he just died. I don't think she's dealt with it really . . . at her age, I'm not really going to bring up those feelings again . . . if that's her way of coping with it.*
>
> (SisterE)

The second coping strategy is *standing up to it*. Before death this covers AFMs' attempts to engage emotionally or assertively with the substance user in the hope of getting them to change their behaviour through realising its impact on those around them. After death we suggest that this way of coping can manifest itself through, for example, interviewees fighting for the good memory of their loved one and not wanting them to be remembered solely as an alcohol or drug user, finding a comfortable balance between good and bad memories and acknowledging the truth about the death (particularly if others were believed to be somehow involved with the death).

> *My sister and I wrote the eulogy together . . . it was our chance just to say to them "This is the kind of person he was" . . . and that we were really, really*

proud of him . . . we were able to put all those things to show who he was really.

(DaughterE)

As time goes on I suppose I remember him more rather than less with the passage of time . . . You know at first I just blot[ted] him out. . . . It was just too painful.

(WifeE)

It was the only thing I could do at that point to honour [my son], I couldn't let it go. It would almost be like saying it didn't matter.

(MotherS)

I wanted it published, I didn't want to hide the fact that [he] died with drugs. Because the way I looked at it, if I could save somebody else, even one, then I think it's all worthwhile.

(FatherS)

The third form of coping is *withdrawing from it*, and before death this covers AFMs removing themselves from the problem or endeavouring to have some independence from it. After death, this might mean the bereaved adult finding their own way to move on with their life, perhaps by responding positively to their bereavement either for themselves or in memory of the one who died, trying to put behind them the death, their grief and the bad memories connected with the substance use (see also Valentine and Walter, 2015; Chapter 4) or using substances themselves (which, for some, might overlap with the *putting up with it* category discussed earlier).

The fact that you could say, well, it's better that they're dead than have somebody that uses in society . . . that really affected me to the point where that's what really made me want to explore drug use further and understand it better and help . . . I was doing a lot of work on drug-related deaths . . . speaking to police and practitioners and people that are still using and as many people as I could . . . to try to gather all these stories together and make this film [to] tackle . . . those kind of stereotypes and ignorance and stuff in society.

(BrotherS)

The first probably three years after he died were just awful, because anything that reminded me of him all it brought back was like these bad memories and stuff . . . whereas it doesn't do that so much now.

(DaughterE)

I was stupid after he died with blocking everything out through all the drug abuse . . . I wasn't interested [in counselling], I just wanted to get out of my face.

(PartnerS)

Though we have stuck with the typology developed from our earlier pro-gramme of research, it is possible that the three broad categories of coping do not so easily apply to AFMs after the person has died. This is because our work has focused on the dilemmas facing AFMs while their loved one is actively using substances. Our data suggest that bereaved AFMs face a new set of coping dilemmas after the substance using person has died, but also dilemmas which are still majorly related to having to deal with the substance use and its related effects, such as managing feelings and emotions, communicating with others about how the person died and remembering the deceased, and the stresses and strains associated with them and their life. Our view is that such dilemmas of coping can persist after the person has died, but it is possible that our typology needs further adaptation to better reflect coping following such a bereavement.

Support

As has been discussed elsewhere (Chapter 6; Templeton et al., 2016a) our inter-viewees painted a very mixed picture of the support which was offered to them or which they accessed after death and of what they found helpful or unhelpful in the response of others, both formally and informally. These experiences lead us to make three suggestions related to the support part of our adapted SSCS model for those bereaved through substance use.

First is the response of authorities and officials, particularly the police, coro-ner and coroner's officers and the procurator fiscal. A poor response which lacked compassion or consideration for the bereaved, or which lacked infor-mation about the necessity of following official protocol, could increase stress (Chapter 2).

> The police basically just said, "We've found your daughter, she's dead, 'phone that number the morra [tomorrow]" and left.
>
> (MotherS)

> The coroner was on another planet really – as in he was sitting there doing his job and it was almost like watching a play because he had a script he was following really.
>
> (FatherE)

On the other hand a more considered approach, where officials explained what was going on or showed some empathy with or compassion for the bereaved, could be comforting at this incredibly difficult time.

> [The coroner] explained how long the post-mortem would take and when we would get the results and that he would ring me and that the best thing I could do would be to just stay at home and not do anything until that process had gone through. And that was helpful, because otherwise I might have gone into a complete flap unnecessarily.
>
> (DaughterE)

And the whole time the police stayed with me, they wouldn't go away . . . I said you must be busy and they said no we'll stay, I said would you like some tea, no would you like some tea, that's what they asked me . . . some time later one of the WPCs came back to the house . . . it was about three weeks after the event. . . [she] very kindly asked after, she said well how are you all coping? And how's your daughter? So the Police do give you a bit of support.

(ParentsE)

Second, is the response of other practitioners, including, for example, funeral directors and specific bereavement counselling services which interviewees accessed or tried to access.

I went to see a bereavement counsellor, I spent about a year working with her. . . [it] was enormously helpful . . . I think that was the first time in my life when I was helped to realise it wasn't my fault that she drank.

(DaughterE)

[The bereavement counsellor] asked me what had happened and I said what had happened to [my son] and he said to me I don't know much about drugs. And I said to him but I am not here about drugs, I am here about loss, but he just didn't seem to understand . . . I never went back, I just thought he stigmatised me right away because of the drugs.

(MotherS)

Third, a number of interviewees talked about the informal support which they received from the friends of the deceased's (including, for some, drug using peers), who could fill in gaps in their relative's life and confirm how much they were loved.

When I got to the crematorium I was absolutely blown away by how many people were there. There was probably a good 200 people there . . . I was so proud that they loved my boy so much. As a mum my heart was bursting because the love that came from those young people was extraordinary . . . it was just all these lovely young people all relating stories about him and laughing and joking.

(MotherE)

Diversity

We have explored elsewhere (Chapter 5) how aspects of interviewees' bereavement may be different for, or more prominent for, a number of sub-groups within our sample. Considering our findings in this way suggests the importance of the relationship of the bereaved person to the person who died, whether the death involved alcohol or drugs, where the bereaved lives and whether the bereaved themselves experience problems with substances. Similarly, although not prominent in our study, other forms of diversity must also be considered, including ethnicity, which illegal drug[s] are involved and sexual or gender identity. As such, diversity should be included within each component of the SSCS model.

In summary, the data indicate that the SSCS model can be applied to adults bereaved through substance use, both to capture their experiences after the death and to understand that these post-death experiences often continue the stress, strain, confusion, coping dilemmas and isolation they often faced when the person was alive. All aspects of our model seem to be relevant. In other words, AFMs feel stress as a result of the death and the bereavement processes which follow, and this stress can result in strain which can take make forms. Additionally, information and understanding, coping, support and (new to our model) stigma all seem to influence (in either direction) the experiences of AFMs after death.

Discussion

This chapter set out to extend our understanding of AFMs' experience when there is an alcohol- or drug-related death by extending a theoretical model – the SSCS model – of how people can be affected by the substance use of a close other to those who are bereaved through substance use.

We first considered how the SSCS model aligns with what our bereaved interviewees said about the impact on them before death of their relative's or friend's substance use. The vast majority of what they said mirrors other UK and international research in this area. However, in describing their experiences in the context of the subsequent death, some additions to our model may be needed. Most notably, in terms of stress and strain, is the sense that many of our interviewees had anticipated or come close to experiencing (through suicide attempts or non-fatal overdoses), the death before it occurred. Related to this, several interviewees described grieving for their relative or friend before they died because of how the hold substances had over them had changed them. We have also added stigma to the stress arm of the model.

We then moved on to think about how the SSCS model could be applied to what our interviewees said about how the subsequent death affected them. Although our preliminary ideas require further investigation, we have demonstrated that all five core elements of the SSCS model are relevant. Hence, the death itself can be very stressful, not least due to direct or perceived stigma. Levels of stress and strain can be influenced positively or negatively by the bereaved's understanding of the substance use, the death (and subsequent official processes), how they cope and the availability and quality of support. We have also shown that AFMs still have major coping dilemmas related to the substance using person, and we have shown that our broad typology of three ways of coping can be applied to the coping dilemmas which AFMs experience when they are bereaved. Nevertheless, we also think that this area of our proposed new version of the SSCS model requires further investigation, both empirically and theoretically, by synthesising with existing theories of coping in bereavement, for example, Nadeau (1998), Stroebe et al. (2000).

One thing which our SSCS analysis has confirmed is that stress, strain, coping and support after death often comprise a natural and inevitable extension of the challenging circumstances faced by the family before the death. Many of the interviewees made clear associations between how difficult things were for them before the death and the impact of the death itself. Furthermore, for many it was important that after the death they processed both what had gone before the death, which often included gaining a better understanding of the person's 'addiction', as well as the death itself, in order better to manage their grief. This mirrors the work of others (e.g., Holland and Neimeyer, 2011), which has suggested the importance of both the 'event story' and the 'back story' in meaning making and processing grief. Alongside this, many interviewees wanted to find some way to remember the person in which alcohol or drug use did not dominate (see Chapter 4). This need to minimise the dissonance between the person the bereaved knew and how the deceased was perceived by others was also highlighted in Wertheimer's study of those bereaved by suicide (2001).

In conclusion, our previous work in developing this model has proposed that:

> The experience of living with a relative with a drinking or drug problem is a very particular experience. It brings together in some combination elements of stress, threat, and even abuse, often simultaneously affecting different family functions and different members of the family. Worry about the loved relative is a core characteristic. It is bad for the health of family members and for the health of the family as a whole. There is no simple name for that kind of experience.
>
> (Orford et al., 2005, p. 117)

The ideas explored in this chapter suggest that this quote could equally apply to the present study's bereaved participants' experiences both before and after the death. For both versions of the SSCS model which we have presented in this chapter, stigma and diversity are central variables which can influence experiences positively or negatively. This new understanding is critical in considering what support is needed by those who are bereaved through substance use. For people affected by a family member's substance use, we used the SSCS model as the basis of a brief intervention, called the 5-Step Method (Copello et al., 2010a, b). Although the 5-Step Method does not consider those bereaved through substance use, the application of the model which we have presented here suggests that such a development might well be possible and could offer a useful addition to existing interventions.

Acknowledgement

We would like to thank our colleague, Professor Jim Orford, for his considered and helpful comments on an advanced draft of this chapter.

Notes

1   For brevity in this chapter we will refer to the group of people affected by someone else's substance use as affected family members (AFMs), even though the range of people affected (and who we are writing about) extends to others as well, including friends.
2   See www.afinetwork.info
3   The combination of difference and commonality has been termed a 'variform-universal' and is further discussed in Chapter 5 in relation to the diversity of substance use bereavement.
4   In recent years the work of AFINet-UK has extended to include AFMs of problem gamblers (e.g.,Velleman et al., 2015).
5   'E' and 'S' refer to interview participants in England and Scotland, respectively.

References

Adfam. (2012). *Challenging Stigma: Tackling the Prejudice Experienced by the Families of Drug and Alcohol Users*. London: Adfam.

Arcidiacono, C., Velleman, R., Procentese, F., Albanesi, C., and Sommantico, M. (2009). Impact and coping in Italian families of drug and alcohol users. *Qualitative Research in Psychology*, 6(4), 260–280.

Arcidiacono, C., Velleman, R., Procentese, F., Berti, P., Albanesi, C., Sommantico, M., and Copello, A. (2010). Italian families living with relatives with alcohol or drugs problems. *Drugs: Education, Prevention and Policy*, 17(6), 659–680.

Barnard, M. (2007). *Drug Addiction and Families*. London: Jessica Kingsley.

Bortolon, C., Signor, L., Moreira, T de C., Figueiró, L., Benchaya, M., Machado, C., Ferigolo, M., and Barros, H. (2016). Family functioning and health issues associated with co-dependency in families of drug users. *Cien Saud Colet*, 21(1), 101–107. doi 10.1590/1413–81232015211.20662014.

Casswell, S., Quan You, R., and Huckle, T. (2011). Alcohol's harm to others: Reduced wellbeing and health status for those with heavy drinkers in their lives. *Addiction*, 106, 1087–1094.

Copello, A., Templeton, L., Orford, J., and Velleman, R. (2010a). The 5-step method: Principles and practice. *Drugs: Education, Prevention and Policy*, 17(S1), 86–99.

Copello, A., Templeton, L., Orford, J., and Velleman, R. (2010b). The 5-step method: Evidence of gains for affected family members. *Drugs: Education, Prevention and Policy*, 17(S1), 100–112.

Copello, A., Templeton, L., and Powell, J. (2010c). The impact of addiction on the family: Estimates of prevalence and costs. *Drugs: Education, Prevention and Policy*, 17(S1), 63–74.

Da Silva, E., Noto, A., and Formigoni, M. (2007). Death by drug overdose: Impact on families. *Journal of Psychoactive Drugs*, 39(3), 301–306.

Esser, M., Gururaj, G., Rao, G., Jernigan, D., Murthy, P., Jayarajan, D., Lakshmanan, S., and Benegal, V. (2016). Harms to adults from others' heavy drinking in five Indian states. *Alcohol and Alcoholism*, 51(2), 177–185.

Fereidouni, Z., Joolaee, S., Fatemi, N., Mirlashari, J., Meshkibaf, M., and Orford, J. (2014). What is like to be the wife of an addicted man in Iran? *Addiction Research and Theory*, 23(2), 99–107.

Holland, J., Currier, J., and Neimeyer, R. (2006). Meaning reconstruction in the first two years of bereavement: The role of sense-making and benefit-finding. *Omega*, 53(3), 175–191.

Holland, J., and Neimeyer, R. (2011). Separation and traumatic distress in prolonged grief: The role of cause of death and relationship to the deceased. *Journal of Psychopathology and Behavioural Assessment*, 33, 254–263.

Lloyd, C. (2010). *Sinning and Sinned Against: The Stigmatisation of Problem Drug Users*. London: UK Drugs Policy Commission.

Nadeau, J.W. (1998). *Families Making Sense of Death*. London: Sage.

Orford, J. (2016). How does the common core to the harm experienced by affected family members vary by relationship, social and cultural factors? In. *Drugs: Education, Prevention and Policy*, 24(1), 9–16.

Orford, J., Copello, A., Velleman, R., and Templeton, L. (2010). Family members affected by a close relative's addiction: The stress-strain-coping-support model. *Drugs, Education, Prevention & Policy*, 17(S1), 36–43.

Orford, J., Natera, G., Copello, A., Atkinson, C., Tiburcio, M., Velleman, R., Crundall, I., Mora, J., Templeton, L., and Walley, G. (2005). *Coping With Alcohol and Drug Problems: The Experiences of Family Members in Three Contrasting Cultures*. London: Taylor and Francis.

Orford, J., Velleman, R., Copello, A., Templeton, L., and Ibanga, A. (2010). The experiences of affected family members: A summary of two decades of qualitative research. *Drugs, Education, Prevention & Policy*, 17(S1), 44–62.

Orford, J., Velleman, R., Natera, G., Templeton, L., and Copello, A. (2013). Addiction in the family is a major but neglected contributor to the global burden of adult ill-health. *Social Science and Medicine*, 78, 70–77.

Philpott, H., and Christie, M. (2008). Coping in male partners of female problem drinkers. *Journal of Substance Use*, 13, 193–203.

Ray, G., Mertens, J., and Weisner, C. (2009). Family members of people with alcohol or drug dependence: Health problems and medical cost compared to family members of people with diabetes and asthma. *Addiction*, 104, 203–214.

Stroebe, M. S., Stroebe, W., and Hansson, R. O. (Eds) (2000). *Handbook of Bereavement: Theory, Research, and Intervention*. Cambridge: Cambridge University Press.

Templeton, L. (2013). The role of the family in substance misuse interventions. In Mistral, W. (Ed.). *Substance Misuse: Emerging Perspectives*. London: Wiley-Blackwell. Ch. 6, pp. 98–117.

Templeton, L., Ford, A., McKell, J., Valentine, C., Walter, T., Velleman, R., Bauld, L., Hay, G., and Hollywood, J. (2016a). Bereavement through substance use: Findings from an interview study with adults in England and Scotland. *Addiction Research & Theory*. Available at: http://dx.doi.org/10.3109/16066359.2016.1153632

Templeton, L., Valentine, C., McKell, J., Ford, A., Velleman, R., Walter, T., Hay, G., Bauld, L., and Hollywood, J. (2016b). Bereavement following a fatal overdose: the experiences of adults in England and Scotland. *Drugs: Education, Prevention & Policy*. Available at: http://dx.doi.org/10.3109/09687637.2015.1127328

Valentine, C., and Walter, T. (2015). Creative responses to a drug or alcohol-related death: A socio-cultural analysis. *Journal of Illness, Crisis and Loss*. Special Edition, 23(4), 310–322.

Velleman, R., Cousins, J., and Orford, J. (2015). The effects of problem gambling on the family. In Bowden-Jones, H. and George, S. (Eds). *A Clinician's Guide to Working With Problem Gamblers*. London: Routledge & The Royal College of Psychiatrists.

Velleman, R., and Templeton, L. (2003). Alcohol, drugs and the family: Results from a long-running research programme within the UK. *European Addiction Research*, 9, 103–112.

Walter, T., Ford, A., Templeton, L., Valentine, C., and Velleman, R. (2015). Compassion or stigma? How adults bereaved by alcohol or drugs experience services. *Health and Social Care in the Community*. Online version. doi 10.1111/hsc.12273

Wertheimer, A. (2001). *A Special Scar: The Experiences of People Bereaved by Suicide*. 2nd Edition. Hove: Routledge.

Wiseman, J. (1991). *The Other Half: Wives of Alcoholics and Their Social-Psychological Situation*. New York: Aldine de Gruyter.

Chapter 2

The impact of a substance-related death[1]

Allison Ford, Jennifer McKell, Lorna Templeton and Christine Valentine

Introduction

This chapter explores the trauma and shock that bereaved people experience following a substance-related death (Feigelman et al., 2012). We draw on new data about the impact on these bereaved people of services dealing with this kind of death and/or bereavement. In the immediate aftermath of any death, survivors are at their most vulnerable. Yet many of our interviewees reported being faced with the distressing, sometimes traumatic, unexplained/suspicious and complex circumstances in which the person died. As a result they had to negotiate a vast and daunting array of 'front-line' agencies and official processes and procedures, which compounded their distress at an already difficult time (Valentine and Bauld, 2016). These agencies included (i) those whose focus was on the deceased and how they died, such as the police, the coroner (or procurator fiscal in Scotland), the pathologist and, not uncommonly, the media; and (ii) those whose focus was on the bereaved person, such as clergy and family and bereavement support services (see Figure 6.1). The work of the undertakers was uniquely focused on both deceased and bereaved. Yet as we learned from our interviewees, finding adequate support from agencies, both in the immediate aftermath of the death and further down the line, was often difficult. Finding support could be affected by the extent to which the person providing it was able to listen to, understand and respond sensitively to the bereaved's predicament. Poor responses, not least from professionals, only served to exacerbate grief. This chapter therefore focuses on three things which contributed to the impact of a substance-related death: circumstances of the death, navigating the system and finding support for grief.

Circumstances of the death[i]

Having lived with a family member's substance use, in some instances with previous experience of overdose or suicide attempts, some interviewees felt that they had already lost that person (Oreo and Ozgul, 2007). What is more, they had lived, sometimes for many years, with the possibility and/or expectation that the person would die as a result of their substance use (see Chapter 1).

There was really no hope any more, he was definitely dying and it would just be a matter of time.

(MotherS)[2]

In some cases, more commonly deaths involving alcohol, interviewees had time to prepare (for example, because the person had been ill for some time or was in hospital) and hence to be with them or say goodbye. However, this did not always help.

I think although we were expecting it, it didn't make it any easier, I think from what I can remember [we were] just completely emotional wrecks to be honest. I mean because people say if you are expecting somebody to die it might make it slightly easier but it didn't, because he was like so young as well. And watching him over that time pretty much destroy himself it didn't make it any easier.

(DaughterS)

Where a prolonged period of alcohol use had damaged relationships, many interviewees had to deal with complex feelings after the death, such as relief or guilt.

We were all sat round his hospital [bed] and saying goodbye and, you know, even then I felt so angry with him as I always did whenever I saw him, although I never responded in an angry way. I was always kind to him, but . . . I really wanted to lean over there and then and whisper in his ear "you are one fucking bastard, I hate you", but I didn't do it, that's what I wanted.

(DaughterE)

He spent his last pound on putting petrol in the car, attached a hosepipe and that was that . . . it was devastating . . . absolutely devastating and I didn't miss him, I didn't feel guilty, I felt guilty for not missing him, but I didn't feel guilty in any other way. I just thought "thank god it's over". I mourned, but I didn't mourn for him I mourned for . . . like the nice times we had . . . holidays and the fun times when the kids were small . . . but then I'd remember what he looked like the last six months before he died and he just looked horrendous.

(WifeE)

In other cases, more common with deaths involving illegal drugs, the death was sudden and sometimes violent. Some interviewees faced the additional distress of knowing that their loved one had died alone, was not found for some time or was (so the interviewee believed) the victim of murder or manslaughter.

The body had been removed but there was horrible evidence . . . they had to burst the door open and his bedroom door was lying wide open and his bed was all covered in blood and the top of the bed was just blood right through pillow cases onto the floor and the house had been ransacked. And it had been left like that as if they had just come in, removed him you know and it had been left like that.

(Ex-partnerS)

He was wrapped in a carpet and put under a bed and left there to rot for three weeks.

(MotherS)

Some interviewees, usually parents where a child had overdosed, discovered the deceased, usually at home, an experience they found extremely traumatic, in some cases involving attempts to resuscitate the person.

The thought of him being there so long and no one having found him was quite hard.

(SisterE)

He died lying on my settee in the living room. I will never forget that day when I went in and found him.

(MotherS)

I phoned 999. . . and the woman told me how to do mouth to mouth resuscitation and everything, and I kept doing that until the ambulance came.

(MotherE)

The death was especially difficult for those who not long before had watched the person make progress with addressing their substance use. These interviewees faced added shock and disappointment when they discovered that the person was using substances again.

We'd been telling everybody "he's clean, he's not on heroin". Everybody was in the house, the police and all that and they are all looking – they all knew different . . . It made it worse for [wife] to know that he was back on the heroin.

(FatherS)

He wanted to make a clean break of it and he had been off drugs for six months. And he was actually looking really good you know his face had filled out . . . I said to him "[son] you look really, really super you know, I am really proud of you" So that was quite traumatic you know, to see him so well you know . . . and he was ok and then that happened.

(MotherS)

Others' presence at or implication in the death could add to the recollection of trauma, especially evident in drug-related deaths; others' presence at the death often provoked considerable emotion for interviewees, raising questions and causing speculation about their role in the death: Had they been involved?

Were they accountable? Had they done all they could? There were instances where a friend had left the scene, apparently concerned about being implicated in the death or being associated with illegal activity, sometimes after calling the emergency services; interviewees questioned whether the friend could have raised the alarm sooner or did enough to help. Friends removing evidence of drug taking from the scene could also hinder interviewees finding out the truth of what happened.

> *Somebody rang the emergency services . . . this lad . . . He left [son] in the toilet unconscious, and then rang, I guess because he was worried about getting into trouble.*
>
> (MotherE)

> *We don't know too many details about exactly what happened . . . there was only one other person with him at the time . . . a girl . . . we believe, he gave the girl money to go and buy the drugs . . . we believe she then went and bought the drugs, came back and as he was right handed and the needle marks were in his right arm, so we believe that she injected him . . . we believe that she was complicit in that . . . Whether he overdosed as soon as she was injecting him and then whether then she got rid of the drugs, first, before she called the ambulance, we'll never ever know, we'll never know that . . . we still don't know really, how culpable she was.*
>
> (BrotherS)

Some interviewees expressed anger and resentment towards those who were implicated or present at the death because they did not cooperate with the police when questioned about events or because interviewees simply did not believe their accounts.

> *And we'll never know, the girl won't say anything. She has been interviewed by the police twice under caution and she just says "no comment".*
>
> (MotherS)

Seeing these individuals in the community could be especially difficult.

> *I knew the person that actually gave him [heroin] . . . I have [seen them] a few times . . . it's not easy.*
>
> (MotherS)

> *It got to the stage, I was carrying a hammer about because if I seen this woman and even to this day, if I seen her I would kill her. I mean it, she . . . gave [daughter] the methadone.*
>
> (MotherS)

This section has outlined how the circumstances of the death could make the loss more traumatic. Next, we show how the official processes and professionals encountered after the death could exacerbate interviewees' distress.

Navigating the system

In the aftermath of the death, those left behind often had to encounter an unfamiliar and confusing system. In about three-quarters of cases there was a post-mortem, often followed by an inquest in England or investigation by the procurator fiscal in Scotland. In nearly half of cases there was some level of police investigation. Against a backdrop of official processes and procedures, many interviewees were, or perceived themselves as being, on the receiving end of insensitive, judgmental and abrupt responses from the officials and professionals they encountered. Statutory procedures focusing on the deceased, such as establishing the cause of death, involved the police and the coroner (in England) or procurator fiscal (in Scotland) and the pathologist. In addition, newspaper reporters, who expect to report such deaths, were involved in making the circumstances of the deceased's life and death public. Where drug use was implicated, establishing the cause of death required official investigation. The family home could be treated as a crime scene, the deceased's body and possessions taken into custody and the funeral delayed until after an inquest or ongoing police investigation. Such delays created considerable uncertainty for the family, who may have felt under suspicion as well as deprived of the person's remains.

In general, interviewees were resigned to these being necessary processes which, apart from the funeral, they had little influence over or detailed involvement in. Some highlighted that having to endure a police investigation or inquest went against their wishes to keep the death a private matter. It was often stressful waiting for responses from officials. Interviewees were often told that officials "will be in touch", rather than provided with detailed explanations of processes and procedures and helpful points of contact. People tended to be in contact with officials when they were "numb" with grief and unable to ask necessary questions. Families often expected the post-mortem and police investigations to provide answers and closure. Yet they only provided disappointment and/or shock if the outcome was not what was expected, leaving family members with only more unanswered questions around the circumstances of death.

Establishing the cause of death: post-mortem and death certificate

For many, the immediate aftermath of the death was a blur. Recollections of this time primarily focused on the delay the post-mortem caused the funeral, typically one to two weeks. In extreme cases, the wait was as long as eight weeks, with families confused over why the process was so protracted. This wait was difficult, with interviewees unsure of what to do in that time. Without a death certificate, they were unable to start making arrangements or perform administrative tasks regarding the death. However, a few interviewees had been provided with an interim death certificate while investigations were ongoing. They regarded this as useful, enabling them to engage with the necessary agencies. For one mother, anxious about others' responses, this gave her some relief that the true cause of death did not have to be disclosed.

> There would be an interim cause of death put on a death certificate which was 'unexplained' or. . . 'not yet known' or whatever it is. So we got an interim death certificate and we were able to register his death and have the funeral . . .You know it's hard to take something that said methadone and alcohol around various places that needed evidence of his death . . . I just didn't want anybody dismissing my son as a druggie or something you know . . . so yes, I mean firstly it helped in order that we could have the funeral and do everything and secondly it helped for that, you know I had to go to the bank with his death certificate to close his accounts and so it felt a wee bit easier somehow.
>
> (MotherS)

After the post-mortem, interviewees recalled further waits for pathology or toxicology reports. The timing of receiving this information could have considerable impact on them. Reports often contained difficult information for the bereaved to hear, sometimes resulting in further questions. It could feel like a setback when they felt they were making progress. In some cases they had a very long wait, even months, to receive these reports, with no updates provided.

> It took three months before we got the autopsy report back and they told us it was a methadone overdose . . . and they said that there was no alcohol and no valium which I found really strange because I knew she had been taking them. But none of them were in her system, just the methadone . . . I was shocked, that was the last thing, it had never even crossed my mind. Why would she take methadone?
>
> (MotherS)

During this stage the bereaved were often uncertain if they were entitled to access documents and view the body. Some interviewees were proactive in chasing pathology reports. Some found out much later that they were entitled, for example, to view the body, causing them additional upset that through not having been informed, they had missed such an opportunity.

> I didn't [view the body]. None of this was really offered to me or occurred to me.
>
> (DaughterE)

Interviewees' beliefs about the cause of death could also be at odds with the official report provided by medical experts. Many told us that details they considered to play a key role in the death were excluded from toxicology or pathology reports. There were mixed experiences of the specific inclusion, or exclusion, of substance use on the death certificate, which interviewees found helpful or unhelpful depending on their perspective.

Didn't have an inquest because they said that he died from aspiration pneumonia, there's nothing on [death certificate] about his drug use. And that makes me annoyed as well because like that will be completely missed, he will be a statistic that is missed for the drug.

(MotherE)

The death certificate which her Mum sent me . . . said that toxicology was clear and I thought hang on a minute she's in a nightclub and drinking, this doesn't add up, however they were so pleased at that . . . you know if it makes it easier for them to comprehend that's fine . . . but for me . . . I'd want the truth really and I sort of needed it.

(PartnerE)

Investigations: police involvement, coroner (England) and procurator fiscal (Scotland)

Police involvement in the aftermath of the death tended only to increase interviewees' distress. In dealings with the police, many interviewees felt stigmatised or believed that the death had not been properly investigated. Some felt looked down on by officers or that they were viewed or questioned as criminals during investigations. A liaison officer was provided in only a few cases.

There were many negative responses from police, such as having no specific point of contact, being transferred around different people when making enquiries, feeling that details were being kept from them, half-hearted approaches to see how the interviewee was coping, dismissive responses to any suggestions for lines of enquiry, feeling that they were not allowed to contact the police and being perceived as too demanding or a nuisance. In a small number of cases interviewees had made formal complaints about police handling of the situation.

We got home and the police came . . . I just felt as though they were, you know it was a drugs death, you didn't matter.

(MotherS)

I had a telephone number that the police that came to visit me gave me. If I wanted any information to ring . . . so I rang him . . . he was awful. He told me off, told me I shouldn't have been ringing that number and not to ring the number again, he was just horrible.

(MotherE)

There were some particularly negative accounts of police officers informing next of kin about the death despite the significance of the moment. In some cases, interviewees had not been given adequate information; for example, there were several accounts of officers being unable to say where the body was being held. There were also reports of the police failing to inform interviewees about the death; they had either found out about the death from other sources or there had been a delay in police informing of the death. In one case the police had got the deceased's name wrong; in another, an interviewee's mother had been told about the death in her workplace, a shop where customers were present at the time.

They just came in, stood in the middle of the room, told me, stayed for two minutes and then went.

(MotherE)

The police weren't brilliant, the way they told my mum. She has a post office [they told her] in the shop with customers there.

(SisterE)

The police questioned some interviewees in the immediate aftermath of the death at a time when they were most distraught and in shock. This was extremely difficult, particularly when officers' questions seemed meaningless, inappropriate or repetitive.

How do [the police] expect me to answer questions when I've just been told my son has died . . . it doesn't matter what kind of person she is or what kind of person he was, you try and show a bit of compassion.

(MotherS)

At the scene of death, police officers could take over private spaces. Where the death happened at home, many different people would arrive to investigate the scene, cordoning off rooms and denying family members access. In such cases, interviewees felt confused about what was happening around them.

The police arrived. They were uniformed police and detectives . . . We stayed in our front room and people were coming and going and we didn't really know [what was happening].

(MotherE)

And there seemed to be a lot of policemen and they said "we're searching the house". I said, "Have you got a search warrant?" He said "we don't need a search warrant". I questioned, you know, how they could do this, but they were very sort of – they treated us terribly.

(MotherE)

The deceased's belongings were especially important to interviewees and an area of particular frustration for interviewees' interactions with the police. There were several accounts where interviewees regularly contacted and chased police for items the deceased had on their person at the time of death or items taken by police to help with inquiries. There could be uncertainty about where items were, with police responses to enquiries appearing unhelpful, and an apparent lack of procedure to return belongings. For some, the person's belongings were never returned despite being told they contained nothing useful for the police. In other instances, where interviewees came to collect belongings, the process was described as cold, brisk and sterile, with items packed in clear plastic bags and no recognition that this could cause distress.

> They took his phone and . . . I never got his phone back. I tried to contact somebody over a period of time and you know you just get passed from pillar to post . . . There were photos on his phone that I wanted to see you know, because I remember one day he was standing there and I thought "you know you are looking great", he was a nice looking boy . . . I took a picture of him so that I would have it because I knew he was going to die sometime . . . I thought "I will be able to look at this when you die".
> (MotherS)

> I was really concerned because I hadn't got [partner's] computer and his phone, his passport, his wallet, his . . . all his stuff, I haven't got anything, I wasn't allowed to take anything out of his room and of course, not having [him], not having any of his stuff, I was getting myself in a bit of a tizz . . . anyway eventually, [police officer] phoned up and said, look come down and get [his] stuff . . . Anyway, picked all the stuff up, and it's all in plastic bags and all labelled and it's all so sterile and, oh it was just horrible. Brought it back.
> (PartnerE)

Where experiences with police were good, these were characterised by continuity of care, regular updates on investigation developments and consideration of the deceased's situation. Positive examples included non-judgmental responses and acknowledgement that the death and its investigation were difficult. Interviewees especially appreciated officers who appeared genuinely sympathetic, spent time with them and undertook small acts of kindness, such as taking the interviewee home after being informed of the death, making cups of tea, giving a hug, saying "I'm sorry" and staying until the body was removed.

> They [detectives] were very nice. I didn't feel they were being judgmental or anything towards me or [him], they were very sorry about [his] death.
> (MotherS)

As with the police, the approach of the coroner or procurator fiscal (PF) (and their staff) was important. Some interviewees experienced a lack of support or at least empathy and understanding from these officials, describing them as distant and lacking in compassion and warmth. Others found them sympathetic, comforting and respectful.

Her manner was terrible, there was no warmth in her. And I felt as if we were the wrong ones.

(MotherS)

The inquest was incredibly professionally and sensitively done . . . it was conducted by a woman who was gentle, sensitive, unrushed.

(MotherE)

In Scotland there was a lack of consistency in communication with the procurator fiscal, whether face-to-face meetings, phone calls or letters. Some interviewees had no contact at all despite the initiation of an investigation; others had initial contact but then never heard the outcome and whether anybody implicated in the death was charged. There was sometimes uncertainty about whether the bereaved person was entitled to contact the procurator fiscal or arrange a meeting to discuss the situation.

Interviewees felt that decisions made by these agencies were beyond their control. It was particularly difficult for them when no charges were brought against others implicated in the death. On occasion interviewees heard new information at the inquest, which could be upsetting and difficult.

The coroner service was established to inform the state of who died, when, where and how. The bereaved family were informed of the coroner's conclusions, if at all, only in passing, and it was up to individual coroners whether or not they saw their role as including that of sensitively informing the family. Some of our sample's deaths occurred during a period when the English coroner service was coming under widespread criticism for its handling of family members (Davis et al., 2002), resulting in a new Coroners and Justice Act 2009 which aimed to "put the needs of bereaved people at the heart of the coroner service". It is therefore not surprising that our interviewees had mixed experiences of coroners, coroner's officers and inquests.

Death reporting and the media[3]

The death was reported in the media (usually the local press) in almost one-quarter of cases, reflecting how death from illegal drugs or excessive alcohol attracts widespread public attention and debate. Media reporting focused mainly on drug-related deaths (Guy, 2004), although accidental deaths where alcohol was involved could also attract attention. Interviewees described two phases of reporting: initially, when the death occurred, to report the event itself; and later, at any subsequent inquest, to report the verdict of the cause of death. Media intrusion was therefore experienced at particularly distressing times.

Those who described a lack of choice or control found media attention much more difficult to cope with than those who felt more involved in the process. Some interviewees who were directly approached or saw reporters making preparations for newspaper articles either asked the reporter or contacted the newspaper editor to ask for the death not to be reported. Often, reporters told individuals that the media had a duty to report such deaths. Interviewees unable or unwilling to engage with the media recalled more negative experiences.

> *Dreadful. They are so intrusive . . . They approached us. We said "Absolutely not . . . We're not talking about it at all. We don't want any photographs; we don't want anything". They visited all our neighbours and asked them for comments.*
>
> (MotherE)

> *I just didn't want to go into a newsagent and see [son's] face. I couldn't bear looking at it without bursting into tears . . . I just didn't need that. So I said to [editor] "do not put it in the paper" . . . . After the inquest they put everything in, everything. And even [family member] didn't know anything . . . of course then found out that it was her fault apparently . . . So consequently, she then went into a big depression and felt awful and dreadful.*
>
> (MotherE)

Making the death public through press reporting threw up difficult issues for interviewees. Some were distressed that newspaper accounts presented a different version of events than the one they wanted to believe or make public. In the example earlier, press coverage revealed to a family member details that others were trying to protect her from, resulting in her feeling responsible for the death. In the next example, the family were trying to protect their son's memory by telling people his death was the result of an accident.

> *If they hadn't bloody published it . . . we would still be in that, "it was an accident", that's all we told people . . . Nobody knew about the situation, how he'd gone . . . But now, because they've put it in the paper . . . They said he was found with a bag on his head, they said he suffocated to death . . . it was unnecessary; they didn't have to put it in. I mean all the people, already bloody knew he wasn't there anymore . . . they didn't have to know all the grotty details.*
>
> (MotherE)

Another father described how the thought of the local community reading about his son caused him a great deal of anxiety.

> *They are not just reading about any Joe Blogs you are reading about my son who has died. That kind of weighed down in my head a wee bit. But again there is nothing much I could do about it . . . It added to the sadness.*
>
> (FatherS)

One mother was so fearful that a verdict of drunk driving would be made public after the inquest that she hired legal assistance in an attempt to stop media reporting.

I decided to pay a barrister as well just in case . . . somebody to come along should there be press there to deal with that for me and with the issues . . . I wanted [to do] everything I could do to not blacken my son's name.

(MotherE)

In contrast, others opted to engage with reporters to try and influence how the death, and the person, was framed in the reporting.

It's horrible. There they are, you know, like vultures in the coroner's court waiting. I actually spoke to one young lady from the [local paper] and I said, you know, "I hope you don't have to use this story, I'd rather you didn't because I said it's my son and it's so distressing" and she just looked at me and she said, "Well it's my job" . . . so in-between that and the story coming out . . . I had the presence of mind then to phone up the [local paper] and say, "Look, can I speak to the editor?" . . . she was very good and said, "Look, we have to report it because it's our duty but at the same token if you'd like us to do a little personal piece about him" and that was very good of them. So it softened it slightly.

(MotherE)

[Son] died on Sunday. It was either the Monday or the Tuesday . . . the local media . . . on the doorstep. It was my wife that answered the door, it was the usual. . . "we've had reports that your son has died" and [wife] says "please I am not wanting to speak to the papers and all the rest of it, it's a private matter". And the way they got around it was they said "well you've got two choices, you either talk to us or you don't talk to us, we are writing something about it whether you like it or not". So my wife and I made a decision there and then to speak to the newspapers or at least try and control it somehow and say what we wanted to say about it . . . they says if we talked they were going to put our words in, nothing is going to get twisted and they will work with us and to be fair to them, they did.

(FatherS)

Generally, when interviewees worked with the press and had some control, although this could be difficult, it led to better outcomes and less distress. Indeed one mother, who was very proactive in her attempt to influence reporting of her daughter's death, recalled the media's attention as a positive experience.

A friend who works for the media had said to us "what you need to do is write out a report for the reporters, and then give it to them. And then it's up to them if they use it or not" . . . But anyway, they did use it. So it was actually quite amazing, really . . . it was very generous – very lovely.

(MotherE)

The funeral[4]

The complicated, confusing and often protracted web of procedures outlined earlier could have a knock-on effect on other processes such as releasing the body and then the funeral. However, arranging the funeral entailed encountering practitioners focused on the family – a welcome shift. Interviewees gave overwhelmingly positive accounts of their engagement with those providing funeral care such as undertakers, clergy and other funeral celebrants. Generally, these people were commended for their approach – sensitive, discrete and respectful of both the deceased and bereaved, while maintaining professionalism. Significantly, funeral personnel are usually the only practitioners paid by the family, and this is reflected in their care and concern for the family. Indeed, apart from counsellors in private practice, they are the only practitioners who will go out of business if they do not show such care and concern.

For many interviewees, being involved in organising elements of the funeral was particularly important. For some, it was perhaps the first 'official' process since the death over which they felt they had control. Interviewees placed themselves as central agents within this process, and accounts were populated with statements such as "I organised. . . " or "I wanted. . . "

The three women planned it, and we invited them . . . people that we wanted to go . . . we sent something formal to invite them and I planned how I wanted it to be and I managed to speak . . . to say all that I wanted.

(MotherE)

I took great pride in [presiding over funeral details], it was a real honour to do that yes and I enjoyed it, yes . . . considering it was for a death, there was just so much life there . . . everybody just had positive things to say . . . so many conversations and support and celebration.

(SonE)

After the often traumatic circumstances of the death and involvement of officials such as police and pathologists, many interviewees looked upon the funeral as an opportunity to celebrate the deceased as a person, rather than as a 'drug addict' or 'alcoholic'.

There was some laughter, there were some nice stories that were told and that made a big difference because you were remembering how good a guy he was so you weren't thinking about the tragic circumstances that led to him dying . . . we were celebrating him as a person and his life.

(BrotherS)

Undertakers and ministers generally provided helpful and supportive guidance through the process of organising the funeral and carrying out the family's wishes.

These people deserve a lot of praise because they do a wonderful job and they are always so nice and willing to do anything that you want.

(MotherS)

How on earth did we [organise] any of this? We had no idea. But the undertakers were wonderful . . . they just organised everything . . . We had no idea who would do this funeral either. They actually knew a minister at the parish church where my mother lived . . . and she was just fantastic . . . she has just been a tower of strength throughout . . . She had spent quite a bit of time with us beforehand on several occasions, "What did we want, what was most important, how did we want it to be, what reflected [son]?" So in the end it was all, it was exactly as we had wanted it to be.

(MotherS)

Perhaps most importantly for this type of death was the non-judgmental attitude of these service providers. The daughter in the next example was particularly anxious about her mother's funeral, which she anticipated would be poorly attended as her mother's alcohol use had lost her many friendships.

He didn't make any assumptions about the situation . . . I felt I needed to prepare them . . . that she had a post-mortem . . . had a fall in the week leading up to her death . . . she was in a bad state. So I felt like I needed to tell him that he shouldn't expect her looking too pretty. And he was, like I say, very empathetic and non-judgmental.

(DaughterE)

In the few cases where interviewees encountered issues with those providing funeral care, this tended to result from stigma. One daughter was upset that she had been told by an undertaker that she would "just want a budget coffin" for her mother. One mother found out that the undertaker had not dressed or embalmed her son due to him having been an intravenous drug user so might pose a risk of infection.

Finding support

Many interviewees talked about coping with and finding support for the emotional, spiritual and social impact of grief, both early on after the death and later. For some, finding support was a priority only after official processes were completed, whereas others needed it straight away if they were to cope with these official processes.

Interviewees received support for their bereavement from various sources. For the most part, this came from other people, including family members and friends; professionals such as general practitioners (GPs) and counsellors, workplace colleagues, spiritualists or mediums and even passing acquaintances. Local or national organisations or community groups dedicated to supporting people affected by substance use or bereavement (or in some areas, both) also helped our interviewees and provided opportunities for them to support others with similar experiences. Additionally, interviewees developed ways to support themselves in their bereavement.

However, interviewees also experienced a lack of support from various sources, including many of those listed earlier, in some cases characterised by responses that were insufficient, insensitive or hurtful.

Family and friends

Those closest to interviewees were a major source of support. Interviewees gained support from spending time with and talking to family members and friends. This was particularly the case where the person who had died was a family member who was also being mourned by other family members. In contrast, those who found that other close family members affected by the death were unable to talk about it could feel deprived of vital support. Different grieving styles between the two parents (Doka and Martin, 2010) can cause marital difficulties after a child has died from any cause (Riches and Dawson, 2000), and some of our interviewees were not immune from this.

> *I feel a lot has been taken away from me as a person . . . and . . . as a couple. Because [husband] is quite quiet . . . he doesn't talk about him and I find that very difficult. I don't know why we don't talk about him.*
>
> (MotherS)

Experiences of friends' responses to the death were also mixed. Some interviewees found that their friends were very supportive, as in the following example where a mother found that it was her and her husband's friends who provided the most support following their son's death.

> *Do you know, in the early days it was our friends who [helped], no agencies. No, you know, no GPs. Friends were what carried us.*
>
> (MotherE)

In other cases, friends struggled to respond supportively to the death, with some friendships subsequently dissolving.

> *We had some friends who we've lost contact with because I think they just can't, they don't know what to do and they can't bear the idea . . . the death of a child.*
>
> (ParentsE)

Professionals and practitioners

One of the main sources of support for bereaved interviewees beyond the immediate aftermath of the death was professionals and practitioners, particularly GPs and counsellors, often dedicated bereavement counsellors. However, a few interviewees also received support from various other practitioners, including psychics and mediums. Their often-mixed experiences of these professional and practitioner groups and the support (or lack of support) they provided are now discussed with reference to bereavement counselling, general practitioners and spiritual practitioners.

Bereavement counselling

For those who had accessed bereavement counselling, whether privately, sometimes through their employer, GP or a voluntary sector organisation, perceptions of its usefulness were mixed. Interviewees frequently experienced long delays in accessing counselling. Some found it very beneficial and credited the counsellor with helping them cope with their grief. Others, however, were unsure of what they had gained from counselling or were clear that it had not been helpful.

> I know the lady, the lady I saw, she was very sweet but I don't know . . . they've got to stick to, almost stick to the lines, they can't veer off and show . . . which is fair enough I suppose . . . In the end I thought, "I don't know what I'm doing this for".
> (MotherE)

> I found it not particularly helpful . . . The man who did it, I don't know, he always gave me the impression of someone who didn't really know anything about grief or grieving.
> (SisterE)

> Anyway, this woman stood up and said "I am very sorry, I am going to have to go out, because I am finding this too upsetting and I am going to have to get a tissue". She obviously found it too much, so that was appalling.
> (MotherE)

It was also clear that some bereavement counsellors struggled to respond appropriately to individuals bereaved through substance use. Some interviewees felt they and/or the person they had lost were stigmatised by counsellors, whose responses suggest to us that they had made certain assumptions or judgements about the death or the deceased.

> I don't know if I was just unlucky with the man I got . . . he said "I don't know much about drugs you know". And I said to him "But I am not here about drugs, I am here about loss".
> (MotherS)

> [I]n one session the counsellor, he asked me if . . . I found it hard to admit that perhaps [son] was using drugs . . . there was an assumption there that I was not facing reality . . . I didn't go back.
> (MotherS)

Additionally, in one of the practitioners' focus groups, a bereaved mother read her daughter's account of being deeply offended by the response of her counsellor to her brother's death.

> [T]his lady did not seem to have had any training with regards to drugs users and certain things she said I found very offensive. As I had told her about the state my brother's flat had been in when he was found . . . she said to me "well you know, these people don't really care when it's got to that point" . . . She may as well have used the terms such as junkies, addicts, druggies, smackheads, etc.
>
> (Focus group participant, England)

General practitioners

Interviewees' experiences of GPs were mixed. In many cases, GPs had a purely functional role, such as referring interviewees to counselling and/or support organisations or prescribing medication. However, in many other cases, the GP was more significant in both positive and negative ways. Some found their GP particularly supportive in the aftermath of the death, to the extent of going out of their way to help. However, many interviewees were unhappy with their GP's response, for example, when the GP failed to acknowledge the circumstances interviewees found themselves in.

> I find it extraordinary that, given that the GP, you know, our GP and [son's] GP and, you know, that health centre, knew so much about us in our family, that nobody, not a single GP thought to pick up the phone and ring us.
>
> (MotherE)

Interviewees also complained of GPs being ineffectual or indifferent. A young woman whose abusive father died an alcohol-related death experienced more support from her school, where she received counselling, than from her GP.

> I'd gone to the GP with my mum and just literally broken down in front of the GP in floods of tears saying, "Look I need help, I need to talk to somebody, I need to untangle everything". Because it was just everything was such a mess . . . and the GP was like, "Oh well we've got like a two month waiting list" . . . that's actually not good enough . . . it's too late by then . . . I would've gone massively off the rails.
>
> (DaughterE)

Spiritualist practitioners

Some bereaved interviewees sought support from psychics or mediums, with some reporting having received answers to questions about the death or about the deceased person. Visiting a psychic could bring some immediate relief as well as some resolution of their experiences.

> I went to a medium . . . She asked me if someone had died of choking before I had even sat down . . . and then it got to the bit where she said you know he was there and he knew that he had been a really bad husband and a bad father and he wanted to, and he was really sorry about that. And I can remember saying to her, tell him too little too late. And the relief of saying that was amazing . . . that was actually quite helpful for me.
>
> (WifeS)

Acquaintances

Acquaintances' responses tended to be experienced as either very good or very bad. Some interviewees experienced unexpected kindness and helpfulness from people who they did not know well. An interviewee who worked as a doctor said that a patient sympathised by sharing their experience of losing a son.

> There was one patient whose son died of a brain tumour . . . twenty two years ago and he said, "Doctor I still cry about it every day and I know you will" and I said, "I think you're probably right".
>
> (ParentsE)

Another developed a close friendship with a new neighbour who had also recently lost her son.

> And she had been very sweet. When we moved in, you know she came round with a bunch of flowers, invited us round for drinks, you know a really nice, sociable woman. She's about ten years older than me and living on her own. And she said to them [friends], "tell them when they come back that the same thing has happened to me and they can come and talk to me anytime about it" . . . So when we got back, she was driving down the road and I was out walking with a friend that was staying with us and she stopped the car and we both . . . got out and we hugged each other.
>
> (MotherE)

In contrast, other acquaintances' responses to the death were characterised by thoughtlessness and insensitivity, causing the interviewee further distress.

> One woman . . . was going about saying to people that she died with a heroin over-dose and what do you expect because the dad's a junkie. And I beat her up.
>
> (MotherS)

Returning to work after the death was considered a form of support for some interviewees in that it structured their days as well as focused their minds, something already noted in the wider bereavement literature (Silverman, 1986). Support for their bereavement in the workplace was forthcoming for some interviewees who experienced colleagues going out of their way to accommodate the impact of the death.

> *I found as well that my colleagues especially, were really supportive and actually quite amazing at putting up with me and talking about him, not just the whole grief and what was going on at that time, but letting me talk about him . . . they were pretty awesome.*
>
> (SisterE)

Yet other interviewees experienced colleagues struggling to respond appropriately or sensitively to their situation.

> *[The doctor] wrote a brilliant letter to my boss explaining what parental bereavement was about and how I would be and how she needed to deal with me, but it didn't really work. She called me to her office and she waved this letter around in my face and said: "This is a bit over the top isn't it?" And I said: "Well actually that's how it is". She just couldn't really deal with it.*
>
> (MotherE)

Support from and for others in groups and organisations

Other major sources of support for interviewees were groups and organisations dedicated to supporting people affected by another's substance use or bereavement and in some cases both. As already indicated, interviewees could draw much comfort and support for their grief in spending time with and talking to others who had experienced a similar bereavement. Having been through similar experiences was seen as one of the key benefits of attending dedicated support groups. Some interviewees, particularly those who were parents, attended groups in their local area, some of which were community based, whereas others were facilitated by national organisations. Those who attended support groups, particularly those in Scotland, had very positive experiences of these groups; in being able to share their loss they developed strong, supportive friendships.

As Chapter 5 will discuss, attending community-based groups featured more in Scottish accounts; in English accounts, attending local groups facilitated by national organisations was more commonplace. Mutual help bereavement groups are based on 'fusing' with others who have had similar experiences of loss (Riches and Dawson, 1996), so newcomers are particularly vulnerable when – as often happens – they discover their experience is not shared by others. If they stay, the group experiences tensions. Or the newcomer may leave or, more rarely, start their own more specialist group (Rock, 1998). All this is found in our data.

Although interviewees generally found support groups beneficial, some found that the nature of the death and their relationship with the person who died prior to death could negatively affect their experience of certain groups. In generic bereavement support groups rather than those solely for substance use bereavement, some interviewees found that they were attending alongside people affected by death due to illness or accident. Tensions could arise when other group members, seeing substance use as self-inflicted, perceived interviewees' bereavement as inferior.

> [Y]ou get the "them and us", because you'll get the mothers who; "well my daughter didn't choose to die did she?"
>
> (MotherE)

Additionally, those whose relationship with the deceased prior to the death was not affected by the person's use of substances felt unsure about attending substance-related bereavement support groups where others had been affected, sometimes for many years, by another's substance use.

> I wasn't actually really sure whether we kind of belonged in it or not, because certainly on the, you know the description you know they talk about if your loved one was estranged or this or that and, the answer is no . . . I've been to the group a couple of times, but I don't feel it's quite right for me.
>
> (ParentsE)

A Scottish interviewee whose son died of an overdose after using another person's methadone prescription said how her son died could cause tensions with members of a substance-related bereavement group.

> I feel wary of telling people because I have such hatred for this girl [woman whose methadone prescription was used] and if I was to find a group like that I might find parents of people like her and how could I then talk about how I feel when they are feeling what they feel about their kids? So I don't fit there either I don't feel.
>
> (MotherS)

Groups and organisations, including treatment agencies, were also a source of support because of the opportunities they provided interviewees to help others experiencing similar bereavements. Some found that by supporting others they coped better with their own bereavement, in some cases both before and after the death.

> Me and [friend] started the bereavement group so it was good because no disrespect or that I didn't want somebody that had lost a granny. I think it was really it was an adult child, all the hopes you had for them and that . . .And it was great channelling my energy there and I think that helped me to cope.
>
> (MotherS)

Additionally for some, helping others was a response to feelings of guilt.

> *I do some work now with a charity in [city] called NACOA which is The National Association of the Children of Alcoholics, chiefly because I feel so guilty that I didn't put my children first really, so it was to try and sort of . . . with the benefit of hindsight, you know I don't know how I let it get to the stage where my son was ill, you know, I think I was so wrapped up in the problem of just trying to pay the bills and trying to keep going . . . I lost all sense of perception.*
>
> (PartnerE)

Some interviewees with current or previous substance use problems found bereavement support in groups that follow the 12-step programme such as Alcoholics Anonymous (AA). One father also saw attendance as a way to address issues relating to an historic problem with alcohol brought to the fore by the loss of his son.

> *You canny deal with the grief, I canny deal with the emotions, so you have to, I have to, for me I know what I need to do, I need to be a twelve stepper myself. I need to be out there helping others, but I canny do that until, I need to sort myself out, and that's what I've done . . . the fact that [son] is never coming back; it was just, it's overwhelming. But I've realised now and I've started the ball rolling to combat it.*
>
> (FatherS)

Self-support

Many interviewees drew on their own resources to address their needs, resulting in a variety of different activities (Valentine and Walter, 2015). In the medium term, some interviewees needed to find out more about substance use, including addiction, as a way to understand the circumstances leading up to the death. These interviewees benefitted from extensive reading on the subject, including searching the internet and even undertaking training courses to improve their knowledge.

> *I've done a lot of research. . . [I] understand things a little bit more now.*
>
> (FatherE)

Some interviewees found creative ways to express their grief. A young filmmaker concentrated his university studies on making a film linked to the substance-related death of his brother.

> *When my brother died, I was just going into my third year which is all self-motivated work . . . writing your own briefs . . . I needed an outlet for [brother's death] so I used my course to do that . . . I ultimately made the film . . . and I think that expression was really helpful actually.*
>
> (BrotherS)

For others, self-support meant continuing their relationship with the deceased (Klass et al., 1996), for example, one mother regularly wrote letters and cards to her son. Some interviewees responded to the death by fundraising for charitable organisations, sometimes in explicit memory of the person who had died. Prayer, meditation and quiet reflection, with or without a religious basis, provided significant sources of support for some interviewees.

> So this tin is now absolutely full of letters that I've written to [him] and I'm still writing to him. And that's the only way that I can kind of hold it together.
>
> (MotherE)

> And probably since I've got clean, I've sort of tuned into things like Reiki, I do Reiki, I do meditation, telling you about all the angel stuff . . . that gave me a lot of peace.
>
> (Ex-partnerS)

> The cathedral is a really, really important place to me. I go there quite a lot . . . I will go and sit there when I'm having a rough day or on her anniversary or on her birthday and I will go and light a candle and there is something about that building, it's so beautiful . . . it's somewhere for me to go and be when I need to think about [sister] . . . so I'm lucky that you know I've got that.
>
> (SisterE)

Concluding remarks

This chapter has illustrated how interviewees experienced and coped with alcohol- or drug-related deaths; the often difficult and distressing circumstances of the death itself; the complex, confusing system through which these deaths were officially processed; and the difficulty of finding support for a poorly understood and often stigmatising bereavement (see Chapter 3). At best, although this was rare, support came from a wide range of sources, giving some hope that communities can be compassionate even after a difficult life has come to a difficult end. Although ongoing support will ideally come mainly from everyday contact with family, friends, neighbours and workmates (Kellehear, 2005), the nature of substance-related death necessitates the immediate involvement of the wide range of practitioners and officials discussed in this chapter, so their approach to family members is crucial. The chapter has presented new findings on the implications for these bereaved people of the responses they receive from professionals and practitioners in the immediate aftermath of these deaths. At a time of extreme vulnerability for those left behind, the responses of those in official positions are likely to have a profound impact for better or worse and with long-lasting effects on those who are grieving such difficult and often traumatic deaths. Thus, abrupt, off-hand responses, which in our sample outweighed those that were kind and reassuring, could be devastating.

As discussed in Part II, professionals and practitioners are themselves up against the pressures and constraints of working in a complex, multi-agency, 'system-less' system. However, the situation was far from hopeless, and some interviewees reported being treated with kindness and consideration. If there is a scandal, it is not that official responses were uniformly poor, but that they were often poor when examples of good practice clearly exist. In a wider social context in which these deaths attract stigma, professionals and practitioners are in a key position and therefore arguably have a duty to counter such stigma (see Chapter 6). We hope that our project's practice guidelines (Chapter 7) offer a starting point for all those who come into contact with adults bereaved by substance use to feel better equipped to offer appropriate and compassionate support.

Notes

1   Elements of this chapter are based on and expand Templeton et al. (2016) Bereavement through substance use: findings from an interview study with adults from England and Scotland. *Addiction Research and Theory*, 24(5), 341–354.
2   'E' and 'S' refer to interviews in England and Scotland, respectively.
3   See Chapter 4.
4   See Chapter 4.

References

Davis, G., Lindsey, R., Seabourne, G., and Griffiths-Baker, J. (2002). *Experiencing Inquests*. London: Home Office Research Study 241.

Doka, K. J., and Martin, T. L. (2010). *Grieving Beyond Gender: Understanding the Ways Men and Women Mourn*. New York: Routledge.

Feigelman, W., Jordan, J. R., McIntosh, J. L., and Feigelman, B. (2012). *Devastating Losses: How Parents Cope With the Death of a Child to Suicide or Drugs*. New York: Springer.

Kellehear, A. (2005). *Compassionate Cities*. London: Routledge.

Klass, D., Silverman, P. R., and Nickman, S. L. (Eds) (1996). *Continuing Bonds: New Understandings of Grief*. Bristol, PA: Taylor & Francis.

Oreo, A., and Ozgul, S. (2007). Grief experiences of parents coping with an adult child with problem substance use. *Addiction Research and Theory*, 15(1), 71–83.

Riches, G., and Dawson, P. (2000). *An Intimate Loneliness: Supporting Bereaved Parents and Siblings*. Buckingham: Open University Press.

Riches, G., and Dawson, P. (1996). Communities of feeling: The culture of bereaved parents. *Mortality*, 1(2), 143–161.

Rock, P. (1998). *After Homicide: Practical and Political Responses to Bereavement*. Oxford: Clarendon.

Silverman, P. R. (1986). *Widow-to-Widow*. New York: Springer.

Templeton, L., Valentine, C., McKell, J., Ford, A., Velleman, R., Walter, T., Hay, G., Bauld, L., and Hollywood, J. (2016). Bereavement through substance use: Findings from an interview study with adults from England and Scotland. *Addiction Research and Theory*, 24(5), 341–354.

Valentine, C., and Bauld, L. (2016). Marginalised deaths and Policy. In Foster, L. and Wood-thorpe, K. (Eds.). *Death and Social Policy*. Basingstoke, New York:. Palgrave Macmillan, Ch. 7, pp. 110–130.

Valentine, C., Bauld, L., and Walter, T. (2016). Bereavement following substance misuse: A disenfranchised grief. *Omega Journal of Death Studies*, 72(4), 283–301.

Valentine, C., and Walter, T. (2015). Creative responses to a drug or alcohol-related death: A socio-cultural analysis. *Journal of Illness, Crisis and Loss*. Special Edition, 23(4), 310–322.

Chapter 3

Managing stigma[1]

Tony Walter and Allison Ford

There are clear theoretical reasons why we would expect substance-related deaths, and those who grieve them, to be stigmatised. In his seminal study *Stigma: Notes on the Management of Spoiled Identity*, Erving Goffman (1963) identified three kinds of stigma: physical stigmas, such as disabilities or deformities; tribal stigmas, such as race or religion; and blemishes of character. In contemporary Western societies, those perceived to be misusing substances are particularly vulnerable primarily to this last kind of stigma. Their character may be deemed faulty or even dangerous – even more than the mentally ill, they may be stereotyped as mad, bad and dangerous to know.

Although liberal academics and professionals may strive to eliminate it, stigma exists for a reason. Crocker et al.'s review paper (1998) identifies a number of functions played by stigma. Looking down on outgroups can enhance both personal self-esteem and collective identity. It can justify social inequality and the dominance of certain groups. And finally, 'terror management' research has found that we tend to become more punitive toward those who challenge cultural norms if we are at the same time reminded of our mortality – and, of course, substance-related deaths both challenge and remind.

Phelan et al. (2008) offer a helpful typology of stigma's functions, compatible with Crocker et al.'s functions and Goffman's three types. First, and this is most likely with physical diseases such as AIDS believed to be infectious, there is health enforcement: stigma keeps people *away*. Second, there is domination and exploitation of outgroups defined by, for example, race, religion or nationality; tribal stigma functions to keep entire groups of people *down*. Third, and most relevant to substance-related deaths, there is norm enforcement: by marking boundaries of acceptable behaviour, stigma keeps people *in*. The logic here was foreshadowed many years previously in sociologist Emile Durkheim's (1938) argument about the social functions of deviance: by identifying and publicly humiliating the outcast, society uses the outcast to reinforce its own values. Here, the perception of drug- or alcohol-related death as self-caused is crucial: a key value of contemporary Western societies, certainly compared to medieval and other more brutal societies, is life itself. Westerners who take or risk their own life for no good purpose, and by association their intimates, are therefore ripe for public humiliation. That said, this tactic may be softening in the UK, as evidenced by the decriminalisation of suicide in the 1950s (Davies, 1975), although health-related stigmas (not least against smoking) may be increasing (Bell et al., 2015).

Kurzban and Leary (2001) probe further by asking whether any of stigma's functions may have an evolutionary basis. With reference to character blemishes, they suggest that humans are more likely to pass on their genes if all members of the group pull their weight; excluding non-cooperators and those lacking resources enhances group survival. From this we may predict that substance users who manifestly contribute to society are unlikely to be stigmatised, which indeed seems to be the case. Winston Churchill drank very heavily; famous people who make major contributions to society and who die as a result of alcohol or drugs are rarely stigmatised; indeed drugs are seen as part of the rock music scene and even part of the creative process.

Even if some forms of stigma have evolutionary roots, this does not justify stigma. Some so-called 'functions' may be dysfunctional hangovers from earlier times when group survival was more precarious. The specific behaviours and attitudes that are socially stigmatised are not set in stone, as is apparent in the aforementioned decriminalisation of suicide and of homosexuality, or − in the other direction − the increased stigma directed at those engaging in potentially deadly behaviours such as smoking, overeating, or world leaders waging war.

Hendriksson (2016) warns against describing those using substances as "stigmatized" because the word reinforces their otherness, that they "are somehow different from the rest of us", worse that they "may start to believe these stereotypes and negative attitudes about themselves, harming their self-esteem and their chance for recovery".. Because stigma is something done to one person by another, Hendriksson recommends we use instead terms such as "prejudice" and "discrimination". Phelan et al. (2008) argue that the concepts of stigma and prejudice, although stemming from different research traditions (sociologist Goffman for stigma, social psychologist Allport for prejudice), have much in common. Both involve "categorization, labelling, stereotyping, negative emotions, interactional discomfort, social rejection and discrimination, status loss and other harmful effects on life chances, as well as stigma management and coping" (Link et al., 2008, p. 11). Prejudice research on race, gender, age and class has highlighted processes of keeping people *down*; stigma research on disfigurement, mental illness and HIV/AIDS has highlighted processes of norm enforcement, keeping people *away* and *in* (Stuber et al., 2008). Stigma research has also highlighted how stigma can be internalised as shame (Walker, 2014), even in the absence of hostile treatment by others. We are reluctant to jettison the concept and word 'stigma', not least because our interviewees often referred to stigma which they themselves felt, and they referred to being stigmatised − implying an action of others against them. Nor are there easy alternatives for the term 'a stigmatised death'. Hendriksson's warning has nevertheless enhanced the care with which we use the words stigma, stigmatised and prejudice.

Sources and forms of stigma

As shown in Table 3.1, our interviewees felt stigmatised by relatives, colleagues, friends, officials and the media.[2]

Direct stigma

After the death, both interviewees and those close to them could experience comments made by others to be hurtful and inappropriate:

> *A couple of days after it happened, my uncle was on his way to my gran's house and the taxi driver said "Oh did you hear about that junkie that overdosed in Charleston, that's at least another one off the streets". And my uncle reacted to that by punching the taxi driver and subsequently getting out of his taxi.*
>
> (BrotherS)

> *She said, "Do you know, I can't believe you made a mistake with your own son". . . . I said "Oh, I didn't make the mistake . . . I'm picking up the pieces". And she said, "Oh, I didn't mean to upset you".*
>
> (MotherE)

Table 3.1 Sources of stigma

Source	Examples
Police	When [son] died . . . the police attitude at the time, I did feel like, do they think I am a criminal, do they think I am worthless, you can't help but have that go through your mind (MotherE)
Media	It was, like, heroin addict killed . . . stabbed to death . . . once people read that they just think oh well, it's only one more gone (BrotherE)
Other professionals or officials	I got the impression that they [health care professionals] just wanted shot of him basically because he was a nuisance . . . he was on a normal ward . . . and he probably did manipulate them . . . I think they just thought, "Oh Christ, just another addict, let's get rid of him" (SisterE)
Relatives	My aunt didn't want to tell anyone how my mum died, she wanted to say that she'd had a heart attack, she's so ashamed (DaughterE)
Others such as friends, work or wider networks	I had a struggle with a lady I've known . . . her son died as an 8 year old of measles . . . she told me "well [losing a child to drugs] it's obviously not the same as losing a child through innocent [means]" (ParentsE)
Wider society	You are a second class citizen (MotherS) You feel like society looks down on you (MotherE)

In some instances the bereaved person would give the perpetrator the benefit of the doubt and assume that the comment was not intended to be hurtful. Stereotypes of substance users are strong and can influence even well-intentioned comments. This mother, for example, acknowledged that those who have not been through it may struggle to understand both living with and grieving an addicted child:

> *It's difficult because unless you have somebody in the family who is an addict . . . if you've not been in that situation, it's hard for you to give an opinion . . . But, you know, people are very judgmental . . . people say to me, "Oh, well, if it were my child, I would do this".*
>
> (MotherE)

Although these examples occurred within the deceased's community, there were also many reports of institutional prejudice. Interviewees felt stigmatised by police or official investigators such as the coroner or procurator fiscal. Stigma could be felt from particular comments or simply by the official's manner:

> *That was worse . . . they [police] had went in there, it's a hostel for drug addicts and prostitutes so basically I feel they went in there that morning and went 'Junkie' and walked back out . . . even the way they came to the door and just so blatant "Well we found your daughter dead".*
>
> (MotherS)

Media reports of the death were often experienced as stigmatising:

> *The newspaper then, that was the next distressing thing. 'Drug Addict Dies In Supermarket Car Park', you know . . . our local paper. I was horrified . . . It was all sensationalism. . . Nothing about the person, just a headline and that's what they do.*
>
> (MotherE)

Assumed/perceived stigma

Rather than being on the receiving end of specific actions or words, many interviewees assumed they were stigmatised. Anticipating what others will think of them or their loved one, they felt judged.

> *"Her son was on drugs", I got that feeling a few times. "He was on drugs, what do you expect?" . . . that was the impression you got. That was the truth of what they were thinking. Whether they were saying so and so to me, they are saying to themselves, "Ha", do you know what I mean, "Another one bites the dust".*
>
> (MotherS)

And I had this terrible sense of "I don't know what people think of me" . . . I used to hate walking through the door of the church because of people judging me. People judging him.

(MotherE)

This can add to stress:

I was worried "What are everybody in my work going to think of me? What are they going to think about my family? And what are they going to think of him?"

(BrotherS)

In these quotes, interviewees pre-empt others' thoughts and responses, denying themselves the opportunity to tell their stories. Negative stereotypes of alcoholics and drug users dominated perceptions.

But when your son is a drug addict there is a huge stigma so you presume everybody must think "he must be a really bad person" your typical drug addict with the hood up robbing old ladies and you know stealing their handbags you know, and they are not all like that.

(MotherS)

One woman, whose former husband had died, was herself a recovering addict. Feeling exposed, judged and worthless as an addict, she did not seek bereavement support:

I didn't hook into anywhere because you think nobody understands you know. Because I think . . . you are just judged as being an addict and "Oh that's just another addict died".

(Ex-wifeS)

Assumed or perceived stigma can arise from interpreting people's facial expressions or body language, or sensing they feel uncomfortable or awkward:

I think we felt that in the accident and emergency, A&E when she was taken in . . . "It's your own doing" sort of thing, you did sense that . . . Mannerisms and body language and things like that, I can't remember what was said you know, but you just got that sixth sense.

(MotherE)

Self-stigma

Whereas direct stigma is attributable to a specific action performed by another, assumed stigma originates within the individual, raising the question of internalised or self-stigma (Stuber et al., 2008). So as well as feeling stigmatised by others, interviewees had to deal with their own stereotypes of a drug addict or alcoholic. A brother pondered this:

> It was in an area where we know a lot of people, so . . . I was worried about what people would think . . . But everyone was very supportive, especially my friends and colleagues at the time, were very supportive but I think, for me, I was self-stigmatising actually. I don't think there actually was a stigma. I think I was self-stigmatising, I think I was making myself feel like there was . . . because there was never any talk, there was never any mention of it.
>
> (BrotherS)

Some interviewees' experiences forced them to challenge their own assumptions about substance use:

> I had all my own sort of prejudices and preconceptions . . . that kind of got in the way because, you know, I hadn't known anybody else with serious drug problems . . . If you said heroin addict, I would have in my mind a stereotypical heroin addict . . . stealing from his grandmother, being incessantly violent, sitting on the street corner, and being generally quite scary . . . I had a whole learning curve, actually. Heroin – people who develop drug problems can be anybody.
>
> (MotherE)

One mother told how this change in attitude affected her own behaviour towards users:

> Before I found out that my daughter was on drugs. I used to say "The state of they people they're a mess" but now I don't do that. If there is somebody at the bus stop who will talk to you and you think they are a drug addict I will speak to them it doesn't bother me.
>
> (MotherS)

Another mother reflected on how to protect her son's reputation without being hypocritical and contributing to the stereotyping of other drug users:

> I am aware now in my attitude of feeling a bit sort of almost hypocritical in thinking "I don't want people thinking my son is a druggie", and by implication I am thinking "Druggies, he is not one of them". So I am aware of something a bit odd there . . . I don't think I really gave it an awful lot of thought beforehand, I certainly do now.
>
> (MotherS)

Stigma is about shame. Feeling embarrassed, ashamed and guilty is commonplace in this type of bereavement:

I was so embarrassed . . . you feel like it's your fault in some way you're responsible, something that you could, you've done something.

(DaughterE)

You think "Oh my God, why is my son, you know, why has my son turned out a drug addict?"You start to blame yourself.

(SisterE)

Parents, children, siblings, partners can all feel they are somehow responsible for their family member's substance use.

Stigma's dynamics

Drawing on Link and Phelan's analysis of stigma, we will now set out how the dynamics of stigma operate after substance-related deaths. The examples in this section illustrate how varied practitioners can be and how prejudice and stigma are not inevitable. Practitioners can choose to treat the bereaved as a unique human being and not as a member of a stigmatised category, and our interviewees were deeply appreciative when this happened.

Stereotyping

Stigma entails labelling and stereotyping, so that all those labelled a particular way (e.g., 'drug addicts') are assumed to be similar. "Categories and stereotypes are often 'automatic' and facilitate 'cognitive efficiency' . . . (They) are used in making split-second judgments and thus appear to be acting pre-consciously" (Link and Phelan, 2001, p. 369). This is relevant to our study for two reasons. First, those who had died were enormously varied in who they were, what substance use meant to them and how they died, as were our interviewees' attachments to them. Yet in the immediate aftermath of sudden death, professionals and officials typically have little, if any, time to get to know either the deceased or their family. Ambulance crew or police called to the family home may know little or nothing of who or what they may find and may employ stereotypes to help make immediate sense of the situation. Respondents such as this mother were upset when they found themselves on the receiving end of incorrect stereotypes:

I do blame the police . . . I think they thought he was just an old tramp or something.
(MotherE)

By contrast, being treated not as a stereotype but as an individual was highly valued. The same mother's experience of the immediate aftermath of her second son's death was much more positive than the first:

There were two policewomen who came and they stayed and they made us tea and they comforted us [cries] ... And [our son] was known to the police as well because he had been an addict, you know, he had been in trouble and that's awful as a mother. You feel like society looks down on you. But I didn't get that sense, no. They couldn't have been more helpful.

(MotherE)

Professional care taken by police or by staff in the coroner's or procurator fiscal's office to get the facts of each unique death straight was also appreciated.

(The woman from) the fatal accident enquiry unit has been great. She checked it out and discovered that no this guy hadn't died in relation to that girl, she told me that and then said that there was still unanswered questions in [our son]'s notes and that she would do her very best to get answers for me.

(MotherS)

A simple shift of language can defuse stereotyping:

The doctor in A&E who signed his death ... he said "This gentleman had died" and that made such a difference to us. We were upset and I thought he wasn't referred to as "This drug addict has died", you know.

(MotherE)

Doctors' letters referring to 'this gentleman' or 'this lady' are a quaint feature of British medicine, according high status to patients regardless of their actual social class, but its use by this emergency unit doctor spoke volumes to the deceased's parents.

Us versus them

To turn labelling and stereotyping into stigma requires separating 'us' from 'them'. He is mentally deranged, I am sane; they are lazy, we are hard-working. "A person *has* cancer, heart disease, or the flu – such a person is one of 'us', a person who just happens to be beset by serious illness. But a person *is* a schizophrenic" (Link and Phelan, 2001, p. 370). Or a drug addict, or an alcoholic. This separation between respectable us and disreputable them can extend to the substance user's family, both before and after death. Kindness, by contrast, implies connection. Ballatt and Campling (2011, p. 9) note that *kind* is etymologically related to *kin*; kindness derives from feeling connected to the other.

All the quotes in the previous section illustrate this. Labelling and stereotyping distance the labeller from the bereaved family, but kindness implies this could be my son, my mother. This is clear in the following appreciation of the coroner, noted by a father who otherwise met with very little kindness:

She said my decision will be based on what I've heard. It's either going to be death by addiction to drugs or accidental death. And she said "I think on the balance of probability it really is accidental death" and she just looked at me and said, "Is that okay?" and I said, "Yes", and she said, "Okay", which was very nice. And she's fine. She was on my side.

(FatherE)

We should note that coronorial kindness over the verdict is not entirely unproblematic. It may bias death certification, and a kind coroner who avoids a suicide verdict may perplex and disturb witnesses who discovered the body.

Feeling with, feeling against

Compassion literally means 'feeling with'. Stigma, by contrast, entails disgust at the other – feeling against. Labelling, stereotyping and separating us from them are cognitive judgments, but they are also inherently emotional judgments communicated through feeling. Thus in this mother's quote, cognitive judgement (just 'another junkie') and emotional disgust ('clearly didn't like him') are connected:

The Police were very much of the opinion that this was yet another junkie. Clearly didn't like him because he had a history.

(MotherS)

A mother and father spoke about the inquest:

But the PC who dealt with it . . . he wouldn't face us. He wouldn't talk to us . . . He didn't come to speak to us.

(ParentsE)

By contrast, a respondent commented on the unexpected positive ambience in the hospital mortuary:

I thought well it's going to be very clinical, I think you watch too many detective movies. It wasn't like that at all . . . He was lying and he looked serene . . . with lighted candles . . . A little bit of ceremony . . . I was glad they did that.

(FatherE)

To conclude this section, despite stigma's social functions, choices can be made by professionals, family, friends, neighbours and workmates whether or not to stigmatise. Although treating someone as a person may, after a stigmatised death, be swimming against the cultural tide or the occupational culture of one's work group, it is not rocket science.

A stigma hierarchy

Our interviewees felt that the degree of stigma attached to substance use depends on the type of substance and the stereotypes attached to it – certain substances have a poorer and more negative image than others. Heroin was felt to attract the worst stigma, associated with the most negative stereotypical image, often represented by the term 'junkie'. Heroin users and their lifestyle are frequently stigmatised as 'dirty', 'filthy' and 'seedy'. Heroin's position at the bottom of the hierarchy of shame was illustrated by a mother who compared heroin with cocaine, which she saw as 'clean' and with a higher class of user:

> And even among the drugs themselves . . . there's a kind of hierarchy. Heroin is always looked on as a dirty drug. Cocaine, while it can be just as damaging, it's seen as quite a clean drug. . . . It always used to be used by the very well off who could afford it.
>
> (MotherE)

The user's perceived lifestyle is therefore a key element in the stereotype. This interviewee's son was addicted to temazepam. His addiction did not prevent him continuing in employment, so his mother did not consider it as bad as heroin addiction:

> He still went to work and that and he used to call anybody who injected 'dirty junkie bs'.
>
> (MotherS)

With heroin, there is evidence of further stigma attached to intravenous use compared with smoking heroin:

> Smoking it somehow seems less desperate than injecting it into any vein that you can find. I guess there's almost like a snobbery about it, it's ridiculous isn't it . . . but somehow it seemed more acceptable.
>
> (MotherE)

When heroin has been involved in the death, interviewees often focused on the deceased's own disdain for intravenous use. The same mother cited this as evidence that her son would not use heroin intravenously:

> Anything to do with heroin, he was abhorrent to it, absolutely abhorrent, hated it, thought it was a filthy, disgusting drug so when he announced to me that they offered him to smoke it and he said, "Oh mum, I did it once, for God's sakes I won't ever do that again, ever" . . . As far as I know, as far as I know he never, ever, ever, ever injected heroin, ever. In fact it frightened him but he felt by smoking it that was acceptable.

Some interviewees focused on others' involvement in the death; believing that someone else would have administered the drug kept their family member out of the stereotype of 'intravenous user'. Perhaps unsurprisingly, parents could be particularly protective in this way of their child's (and their own) reputation; we recall a Scottish study 45 years ago which noted how parents of juvenile delinquents typically blamed peers for leading their child astray (Walter, 1978). Interviewees perceived not only a hierarchy of drugs but also a status difference between using drugs and using alcohol. Those who had experienced a drug-related death often contrasted it with alcohol deaths, feeling alcohol to be different because of its legality and social acceptability. One mother said:

> I don't think there's as much stigma if it's alcohol . . . Alcohol is legal and it's socially acceptable because lots more people can use alcohol as a social thing.
>
> (MotherE)

Another mother described experiencing stigma in a bereavement support group:

> We met a guy who ran a group in Glasgow . . . he wanted to include alcohol in the group, the group was a drugs group but he wanted to include alcohol. And he had the two groups on two separate nights – an alcoholic group and the drug group – but he wanted to bring them together. But the alcoholic group didn't want to go with the drug group. "We are not going with that junkie scum".
>
> (MotherS)

Another mother considered that the care offered to her son would have been better if alcohol rather than drugs had been involved.

> It would never have been so bad had it been drink . . . the effects are the same. The despair is the same. There is just more help. There is more dignity if you are an alcoholic by a long chalk. People treat you like a human being and the hospitals treat you as a human being, but they don't when you are a drug addict, they treat you like you are the scum of the earth.
>
> (MotherS)

However, several interviewees whose relative had died in hospital after a long history of alcohol use spoke of how *undignified* the death was. Many interviewees stressed the importance of dignity at the death, even if there was little dignity, perhaps in an attempt to counteract the stereotype of substance users living an undignified life.

Disenfranchised grief

Doka's (2002a) concept of disenfranchised grief is well established in bereavement research and practice. It can refer to several ways in which grief is not socially acknowledged. For example, when someone's identity is spoiled (Goffman, 1963) due to mental illness or substance use, after death family members may encounter the assumption that the person they loved was not worth loving and not worth grieving, or that they must be relieved that the person who had for years caused them so much trouble can do so no longer. Doka (2002b) argues that stigmatised deaths can disenfranchise grief, especially if the death is perceived to be self-inflicted; many of our interviewees were acutely aware of this.

Many described instances where those around them had made this assumption explicit by comparing a substance use death to a death by other causes such as cancer.

> Sometimes I feel that [if] somebody died in an accident or somebody dies of cancer, there is so much support and everybody thinks that's so terrible, that's awful. But somebody died of drugs, oh well that was their own fault.
>
> (MotherS)

A daughter felt her grief was devalued because others saw her as complicit in her mother's alcoholism:

> I also had a sense that when I went to the funeral that there were certain people that knew about my difficult relationship with my mum but probably only heard her version of it and some of them didn't speak to me and only spoke to my sister and yeah, I felt as though I was to blame for her drinking.
>
> (DaughterE)

Her disenfranchisement was also to an extent self-imposed: she described how her relationship with her mother led her to question whether she had the right to be upset when her mother died. This is further complicated by the relief she felt at the death.

> It took me a while to realise that I had the right to be upset . . . We always argued and fought and I just thought the fact that I always tried to do something, tried to make it better in some way, small way, made me realise that actually I did care a lot about her and that I did have the right to be upset. But it took time for me to get there.

Doka (2002b) highlights two strategies for getting your grief acknowledged after a stigmatised death. Both entail identifying and then spending time with others who acknowledge the grief. One strategy is to identify what he calls 'the own' – those who have been bereaved the same way, typically found in a mutual support group. The other is to identify 'the wise' – people who, although not bereaved in this way, nevertheless understand, such as a counsellor or therapist.

Neither strategy is full proof and may even rebound, creating a new experience of stigma. Chapter 1 described how one mother felt stigmatised by a bereavement counsellor who refused to engage with the role played by drugs in her son's life and death. Mutual help groups are often turned to in bereavement because "here at last are people who have been where I've been" (Walter, 1999). Unfortunately, not all bereavements are the same, and even in a mutual help group, stigmatising distinctions may be experienced. To expand a quote from Chapter 2:

> *I'm in a Compassionate Friends group (for bereaved parents), you've got people who've lost children through cancer . . . you get the 'them and us', because you'll get the mothers who "Well my daughter didn't choose to die did she?" . . . There's one lady who has lost her only daughter to a brain tumour . . . We'll both be supporting each other and crying because we've lost our children but then there will be this, "Yes but mine didn't choose to go".*

> (MotherE)

In his study of family members bereaved by homicide, Paul Rock (1998) shows how members of mutual help bereavement groups fuse or bond through sharing similar experiences, but those whose experience differs from the group norm can find themselves excluded. Each group, although aiming to give everyone the voice society disallows, nevertheless generates its own group understanding of what, for example, losing a child is like. In Rock's study, some parents bereaved by homicide left The Compassionate Friends to form their own group for parents bereaved by homicide. For substance-related bereavement, however, there are few specialised groups in the UK. Those bereaved by suicide, the most obvious cause of self-inflicted – and hence potentially stigmatised – death can join a local SOBS (Survivors of Bereavement by Suicide) group, but there is no organisation for those bereaved through substance use with anything like UK-wide coverage. Why this is, we cannot say for sure, but it may be that those bereaved by suicide bond more readily than those grieving the very diverse deaths involving drugs and alcohol. It has been suggested (de Vries and Rutherford, 2004) that online even those with very minority experiences can find others on the planet with similar experiences. Few of our interviewees, however, found solace from sharing experiences online.

Responding to stigma: family members

Being stigmatised is something either others do to one or one does to oneself. But interviewees spoke about the choices they faced in responding to stigma. Doka (2002b, p. 326) puts it astutely:

> *Stigmatized deaths place survivors in a double bind. If they risk disclosure, they may be perceived differently by others and fail to receive the support they seek. Yet if*

they do not risk disclosure, they deny themselves the possibility of support, and they conceal an important attribute of identity.

Should they disclose and risk being stigmatised? Or not disclose and live with the risk of the cause of death leaking out? We look first at non-disclosure before going on to consider disclosure.

Non-disclosure

Many interviewees said they consciously avoided talking about the role of substance use in the death in order to avoid the anticipated stigma. This reduced the chance of being judged by others or having to confront others' stereotypes of the addict, but could also increase isolation and reduce opportunities for support:

> *You feel reticent to say my son died of a drug overdose because it tells people so much. They think it's telling them everything about that person and it's not, you know.*
>
> (FatherE)

> *I couldn't [speak about the death] for a long time . . . When they die from drugs you think "Oh if I say this, people will think oh she is just off her head – he was a drug addict, he must have been a bad person".*
>
> (MotherS)

> *I think the stigma of the drugs is there even in my work place . . . I don't talk about it, there is only very few people that I would speak to about [my son].*
>
> (MotherS)

Another strategy to avoid stigma and elicit a more sympathetic response is to misrepresent the cause of death:

> *My aunt didn't want to tell anyone how my mum died, she wanted to say that she'd had a heart attack, she's so ashamed.*
>
> (DaughterE)

> *My mother . . . told everybody he had a brain tumour . . . Well, for a brain tumour, you'd get immediate sympathy rather than judgement.*
>
> (ParentsE)

Another strategy is not directly to lie, but to be economical with the truth. Some interviewees described becoming very careful with language, carefully considering the meanings of terms.

I had just said that I had lost a close friend and when I was asked what, I just said a drugs overdose. So you know it could cover a multitude of things. So I wasn't saying he was a drug addict . . . it could have been any reasons. And I suppose when I am saying that I was probably the same as his mother and father . . . a wee bit ashamed of the situation.

(Family friendS)

A couple used the word 'accident' to describe their son's overdose:

A lot of people have said, you know, "Well how did he die?" And we've said, and as far as I'm concerned it's completely truthful, well we don't honestly know. It was accident . . . So we've not come in for the, you know the sort of stigma attached to . . . because he was deeply, deeply ashamed of it.

(FatherE)

This helps not only to avoid stigma but also to protect the memory of their son who, his parents believe, would have been ashamed had people known he had died from an overdose. In another example a mother did not want her son to be labelled a 'drunk driver', fully aware of how this would influence how others perceived him:

And again I never say about the alcohol . . . I see people who have been bereaved . . . You see them interviewed on the television and they say, these terrible dangerous drivers, reckless drivers, drunk drivers . . . My son wasn't like that.

(MotherE)

Non-disclosure can start long before the death. Several interviewees said their relative had felt ashamed of their addiction and would not want people to know. In particular, children and partners of those with alcohol problems continued a behaviour learnt during the person's life in which covering up and concealing the truth was the norm.

But the alcohol side with both my mum and my dad, what I've learned . . . is don't talk. I remember getting told . . . when we went into town if somebody said, "Oh, how's your granddad?" "He's not well". I wasn't allowed to say anything else . . . The same with my mum before things got really bad. "Mum's not well today".

(DaughterE)

Some talked about how over time they came to see benefits in disclosure, sometimes because they came to see substance misuse as an illness and to accept that there was no shame in the death.

Disclosure

> *I've always talked about [son's] drug problem . . . we don't let it die down now because it's there, it's in our life it's part of who we are now and we are certainly never going to shove it under the carpet.*

> (FatherS)

Some interviewees report a strong motivation and desire not just to disclose but also to challenge stigma. They write to newspapers, phone into radio programmes or use the funeral to tell the family's story and/or to counter stereotypes; more privately, they may take a different stance from other family members on telling the 'truth' about the death.

One purpose is to evoke empathy, to help people see how it could be their family and to help people understand the impact of substance-related death.

> *I am proud of my son and who he was, I am sorry for what he suffered and I am not proud that he was a drug addict but I am not going to pretend it didn't happen and I am not going to pretend that it didn't happen to my family because I think it's important that people know that it happens to ordinary families and to good families and to caring mums and dads and to parents who would give everything they've got for their children.*

> (MotherS)

Another purpose is to depict the deceased's humanity, to show how they were loved and are worth grieving (Butler, 2009):

> *I said he's not a drug addict, he wasn't – I said [he] was a wonderful son, brother, uncle, grandson, a very special human being, not just a drug addict, you know, I hated that stigmatised thing.*

> (MotherE)

> *We actually gave the eulogy because we thought do you know what, this is our only time to kind of put the record straight because even at school, at comprehensive school, one of my best friends had said "Oh, you know we all know your dad is a drunk".*

> (DaughterE)

The key strategy to challenge the stereotype of a drug addict or alcoholic is to show that the deceased was not just an alcoholic or drug addict but a unique human being, with positive qualities and a loving family.

> *I felt, well, "I'll have to explain to everyone, that he wasn't a junkie" . . . I can understand why people feel the way they do and why people say certain things because, you know, they maybe know of a lot of incidents that have happened; they*

maybe know somebody's who's been jumped [mugged] because of that or somebody might have broken into their house as a result of trying to get drugs. So I understand people's frustrations . . . but they don't understand the person behind it. They don't understand that's someone's son, daughter, brother, sister, they don't understand that's a person and why that person may have gone to use in the first place.

(BrotherS)

I just read, 'Unemployed man dies of drug overdose' and read down through and it was [our son]. Our address, name and address. 'Living with his parents, 30 years old, room full of drug taking paraphernalia. The parents said that he wasn't getting any help for his condition', and that was that, you know, that was all they could find to write about my son . . . I thought I've got to do something about this so I wrote to the editor of the paper, I explained the type of person [my son] was and I don't think the main point about him was that he was unemployed, there was more to [my son] than an unemployed man.

(FatherE)

This strategy of depicting the deceased as unique and loved is used by a number of campaigning organisations, biographies and videos (Gazit, 2015, Skinner, 2012). The strategy, of course, purveys its own stereotype. Although almost every parent we interviewed loved their deceased child, not every adult child loved their deceased parent, whereas others loved them ambivalently. It is possible that some may feel that propaganda depicting all those who die through alcohol or drugs as loved and worth grieving marginalises their own negative or ambivalent relationship with their deceased family member.

In these examples, interviewees use their own experiences to challenge stereotypes. Others challenge stigma by focusing on the language and terminology surrounding drug addiction. The derogatory term 'junkie', often used to describe a heroin addict, reinforces a negative stereotype. Many interviewees highlighted how offensive they found this word; their response was to challenge its use.

That's a word we don't use, 'junkie' . . . we call them drug addicts . . . You know somebody came on the bus and said "Look at the state of that junkie" and I went "Excuse me, that's somebody's child and that word is horrible. Don't ever use it again". "Who do you think you are talking to?" I said. "I am talking to you", I said, "because you don't know the circumstances of that person".

(MotherS)

I mean if anything comes in the papers like 'junkie' or anything, I write a letter to the paper and say you know I think you should really start using different phraseology, because that's not helpful to people's perception of what's going on you know we are trying to help people not really bring them down.

(MotherS)

It is not just journalists whom some interviewees tried to educate:

> *I wrote to our chief constable and said your police need to be taught how to speak to bereaved parents.*
>
> (MotherE)

> *Well we edit leaflets all together, but I was the main writer for coping with judgmental attitudes and that was good for me writing that, the effect it has on people whose children have died from drugs, alcohol and suicide and all that.*
>
> (MotherE)

Valentine (2015) has described a range of creative ways our interviewees challenged stigma. To give just one example, a brother mentioned in Chapter 2 used the medium of film to undermine the drug addict stereotype:

> *(I used) film to try to gather all these stories together and make this film (to try and) tackle those kind of stereotypes and ignorance and stuff in society.*
>
> (BrotherS)

Responding to stigma: practitioners

Being or feeling stigmatised clearly added to many interviewees' distress; several interviews comprised pleas to be treated with kindness and as an individual, not stigmatised as an assumed member of a stereotyped group. Stigma makes a bad situation worse; by contrast, even small acts of kindness can make a big difference.

Both stigma and kindness entail cognitions and emotions. We now reflect on the cognitive and emotional labour engaged in, particularly by practitioners, as they respond to stigmatised deaths. Because we did not interview practitioners, this section does not draw on quotes. Rather, we reflect on what it takes for not only the bereaved but also others to challenge – or at least implicitly through attitude and action – to undermine stigma. In this, we draw on a range of literature in sociology, psychology, death studies and compassion in health care.

Just as both disclosure and non-disclosure were problematic for family members, so practitioners can face a dilemma in the immediate hours, days and weeks after the death. They are mandated to prevent, confirm or investigate the death – ambulance crew strive to revive the person; doctors pronounce them dead; police, coroner or procurator fiscal investigate the circumstances; and journalists seek a newsworthy angle on the death. Unlike funeral directors (of whom we heard very little criticism), none of these personnel are paid to help or empathise with the bereaved. Further, the practitioner may personally disapprove of the family's lifestyle as much as the deceased's. The emotional labour of empathy risks undermining the instrumental task their job entails, especially if – as is increasingly the case – personnel are operating with limited time and resources. Yet not to treat family members as unique persons who suffer like anyone else risks dehumanising not only the family but also the practitioner. We now try to unpack some elements of this dilemma.

Cognitive labour

Death a time of judgment

It is normal for humans to make judgments of others and for those judgments to be based on assumptions and stereotypes. At no time in the life course are we more prone to make judgments than at the end of life. Elderly people may conduct a life review (Butler, 1963), falling mountaineers and others who have a few seconds before they miraculously survive report their life flashing in front of them (Kellehear, 2014), religions have portrayed death as a time of judgment and eulogies assess the value of the deceased's life to those who survive them (Walter, 2016). Thus, judgments are made in the light of one's own impending death and about those who have died, and mourners judge themselves by the deceased's life and values. Such judgments, however, are usually based less on assumptions and stereotypes than on personal judgment of a unique life.

It is unrealistic to ask practitioners encountering people bereaved after substance use not to make judgments. However, unlike family members, funeral directors or funeral eulogists, practitioners such as emergency medics, ambulance crew or police have very little time or knowledge to make an informed judgment about this particular person or their family. Judgments will inevitably be broad-brush. What our interviewees ask is simply that such judgments be judgments of kinship, not of othering.

Meaning-making

Psychologist Robert Neimeyer has researched in detail how mourners strive to create meaning in the face of loss. Meaning-making is critical after traumatic deaths (Zandvoort, 2012, Neimeyer and Sands, 2011) which disturb mourners' sense of the order of things (Guy and Holloway, 2007); struggling to find meaning in the death can predict complicated grief and a poor bereavement outcome (Davis et al., 2000). Not only mourners, but also practitioners, need to make sense of the traumatic deaths they encounter, although this has barely been researched. We know that a person can feel supported or disturbed by the meaning their own family gives to a death (Nadeau, 1998), but they can also feel supported or disturbed by practitioners' meaning-making. Our interviews show how coroners and their staff can help families understand the death and the life that led to it; how treatment agencies' framing of substance dependency as a disease can ease a family's guilt; how the stereotyped meanings some police give to drug deaths can deeply disturb families; how families try, and often fail, to influence the meanings that media give to the death. All this entails cognitive labour.

Emotional labour

The American sociologist Arlie Hochschild (1983) coined the term 'emotional labour', a major part of some occupational roles. Emotional labour entails two norm-governed tasks: manipulating what you feel, according to feeling rules; and manipulating what you do, according to display rules.

Feeling rules

For practitioners to root instant judgments in kinship requires emotional labour. Msiska et al. (2014) summarise two kinds of emotional labour required of health care workers as they work with patients; workers need to repress emotions such as disgust or frustration and to detach from instinctive identification (Mann, 2005). Being kind to someone who repels you entails emotional labour, as does not being incapacitated by sadness at the misfortune of somebody with whom you identify. What our interviewees valued, however, was not total embrace by the practitioner or hours of their time, but simple acts of kindness. Informing somebody that their son has died takes much the same time whether done with kindness or with indifference. The question is whether such kindness might lead not to burnout but, as Youngson (2010) has argued, to greater job satisfaction.

Others see this in terms not of job satisfaction, but of exploitation. Bolton and Wibberley (2014) analyse the labour required of those who provide homecare to frail elderly people; as well as their formally contracted labour, they also find themselves engaging informally in emotional labour that, although highly skilled, is not contractually recognised (England and Dyck, 2011). Are our interviewees asking a whole range of practitioners to add unpaid, unrecognised emotional labour to their already harsh workloads? Care workers are paid to care, but care is not at the heart of the contracted labour of journalists, coroners and police officers – although as noted earlier, a little goes a long way.

Display rules

Although emotional distance in the face of others' pain is an individual response, it is influenced by group norms (Ballatt and Campling, 2011), not least when a practitioner group is repeatedly exposed to human misery and the terrible consequences of human error. Stoical norms, including black humour (Young, 1995), are likely to prevail among practitioner groups which frequently encounter death, especially sudden, untimely or violent death – norms which may conflict with the more expressive norms that have become increasingly influential among Western publics who now encounter death rarely (Walter, 1999). Our bereaved interviewees are *not* asking practitioners who frequently work with traumatic deaths to be tremendously expressive or revealing of themselves; but our interviewees do tell us that small kindnesses can really help. The sympathetic ambulance crew, like the soldier writing to his deceased comrade's family, understands that their occupational group's necessary stoicism need not preclude small kindnesses to a bereaved family. Emphasising the virtues of tact, courtesy and kindness may sound simplistic, but it is what our interviewees emphasised time and again.

Conclusion

We started this chapter by describing some social and psychological functions that stigma plays, causing us to hypothesise that substance-related deaths, embracing those who mourn as well as those who die, might be strongly stigmatised. Stigma – whether direct, perceived or self-constructed – was indeed a major theme in what our interviewees told us. At the same time, it is striking that not every substance-related death is stigmatised, or not by everyone the interviewee encountered. It is also striking how interviewees exercised their own agency in choosing, for better or for worse, how to respond to stigma and thus developed ways to manage and thus to survive it. And as Part II will reveal, we also found a very clear willingness among the practitioners in our focus groups to develop non-stigmatising practice.

Notes

1   This chapter is based on Walter, T., Ford, A., Templeton, L., Valentine, C., and Velleman, R. (2015) Compassion or stigma? How adults bereaved by alcohol or drugs experience services. *Health & Social Care in the Community*. DOI: 10.1111/hsc.12273
2   'E' and 'S' indicate interviews in England and Scotland, respectively.

References

Ballatt, J., and Campling, P. (2011). *Intelligent Kindness: Reforming the Culture of Healthcare.* London: Royal College of Psychiatrists.

Bell, K., Salmon, A., Bowers, M., Bell, J., and McCullough, L. (2015). Smoking, stigma and tobacco 'denormalization': Further reflections on the use of stigma as a public health tool. *Social Science & Medicine*, 70(6), 795–799.

Bolton, S. C., and Wibberley, G. (2014). Domiciliary care: The formal and informal labour process. *Sociology*, 48(4), 682–697.

Butler, J. (2009). *Frames of War: When Is Life Grievable?* London: Verso.

Butler, R. N. (1963). The life review: An interpretation of reminiscence in the aged. *Psychiatry*, 26, 65–76.

Crocker, J., Major, B., and Steele, C. (1998). Social stigma. In Gilbert, D. T., Fiske, S. T., and Lindzey, G. (Eds.). *The Handbook of Social Psychology*. 4th Edition. Boston, MA: McGraw-Hill, pp. 504–553.

Davies, C. G. (1975). *Permissive Britain: Social Change in the Sixties and Seventies*. London: Pitman.

Davis, C. G., Wortman, C. B., Lehman, D. R., and Silver, R. C. (2000). Searching for meaning in death. *Death Studies*, 24(6), 497–540.

de Vries, B., and Rutherford, J. (2004). Memorializing loved ones on the World Wide Web. *Omega*, 49(1), 5–26.

Doka, K. (Ed.) (2002a). *Disenfranchised Grief: New Directions, Challenges, and Strategies for Practice*. Champaign, IL: Research Press.

Doka, K. (2002b). How we die: Stigmatized death and disenfranchised grief. In Doka, K. (Ed.). *Disenfranchised Grief*. Champaign, IL: Research Press, pp. 323–336.

Durkheim, E. (1938). *The Rules of Sociological Method*. 8th Edition. New York: Free Press.

England, K., and Dyck, I. (2011). Managing the body work of home care. In Twigg, J. (Ed.). *Body Work in Health and Social Care*. Chichester: Wiley-Blackwell, pp. 36–49.

Gazit, Z. (2015). (Social) death is not the end: Resisting social exclusion. *Contemporary Social Science*, 10(3), 310–322.

Goffman, E. (1963). *Stigma: Notes on the Management of Spoiled Identity*. London: Penguin.

Guy, P., and Holloway, M. (2007). Drug-related deaths and the special deaths of late modernity. *Sociology*, 41(1), 83–96.

Hendriksson, M. (2016). *Words Matter, SAMHSA: Substance Abuse & Mental Health Services Administration*, May 16. Available at: http://blog.samhsa.gov/2016/05/16/words-matter/#.V1G7rDUrKUl

Hochschild, A. (1983). *The Managed Heart: Commercialization of Human Feeling*. Berkeley, CA: University of California Press.

Kellehear, A. (2014). *The Inner Life of the Dying Person*. New York: Columbia University Press.

Kurzban, R., and Leary, M. R. (2001). Evolutionary origins of stigmatization: The functions of social exclusion. *Psychological Bulletin*, 127(2), 187–208.

Mann, S. (2005). A health-care model of emotional labour. *Journal of Health Organization and Management*, 19(4/5), 304–317.

Msiska, G., Smith, P., Fawcett, T., and Nyasulu, B. M. (2014). Emotional labour and compassionate care: What's the relationship? *Nurse Education Today*, 34(9), 1246–1252.

Nadeau, J. W. (1998). *Families Making Sense of Death*. London: Sage.

Neimeyer, R. A., and Sands, D. (2011). Meaning reconstruction in bereavement. In Neimeyer, R. A., Harris, D. L., Winokuer, H. R., and Thornton, G. F. (Eds.). *Grief and Bereavement in Contemporary Society: Bridging Research and Practice*. New York: Routledge, pp. 9–22.

Phelan, J. C., Link, B. G., and Dovidio, J. F. (2008). Stigma and prejudice: One animal or two? *Social Science & Medicine*, 67(3), 358–367.

Rock, P. (1998). *After Homicide: Practical and Political Responses to Bereavement*. Oxford: Clarendon.

Skinner, P. (2012). *See You Soon: A Mother's Story of Drugs, Grief and Hope*. Spoonbill Publications.

Stuber, J., Meyer, I., and Link, B. (2008). Stigma, prejudice, discrimination and health. *Social Science & Medicine*, 67(3), 351–357.

Valentine, C., and Walter, T. (2015). Creative responses to a drug or alcohol-related death: A socio-cultural analysis. *Illness, Crisis and Loss*, 23(4), 310–322.

Walker, R. (2014). *The Shame of Poverty*. Oxford: Oxford University Press.

Walter, J. A. (1978). *Sent Away: A Study of Young Offenders in Care*. Aldershot: Gower.

Walter, T. (2016). Judgement, myth and hope in life-centred funerals. *Theology*, 119(4), 253–260.

Walter, T. (1999). *On Bereavement: The Culture of Grief.* Buckingham: Open University Press.

Young, M. (1995). Black humour: Making light of death. *Policing & Society*, 5(2), 151–167.

Youngson, R. (2010). Taking off the armor. *Illness, Crisis & Loss*, 18(1), 79–82.

Zandvoort, A. (2012). Living and laughing in the shadow of death: Complicated grief, trauma and resilience. *British Journal of Psychotherapy Integration*, 9(2), 33–44.

Chapter 4

Remembering a life that involved substance use

Christine Valentine and Lorna Templeton

This chapter draws on insights from previous research on the role of remembering in making sense of and becoming reconciled to the death of someone close. Theories of grief have drawn attention to how bereavement may undermine our identity (see, e.g., Berger and Luckmann, 1967; Parkes, 1988, Seale, 2000), that is, our sense of who we are and our role in life, challenging taken-for-granted beliefs, hopes and dreams, leaving us feeling unsafe, uncertain and anxious about the future. Piecing together and keeping alive memories of the deceased person's life and character play an important role in recovering identities of both the bereaved and deceased person, as well as the relationship between them (Braun and Berg, 1994; Walter, 1996; Neimeyer and Sands, 2011). This process has been conceptualised as 'meaning reconstruction' (Neimeyer, 2001) and often takes the form of 'storying' grief, in that telling our story may help to create coherence and meaning in an experience that has brought pain and chaos into our lives (Gunaratnam and Oliviere, 2009). Storying has been found to benefit from two interdependent resources: having one's bereavement acknowledged by others, both publicly at the funeral and more privately through talking about and sharing one's grief with sympathetic others (Walter, 1996, p. 2); and taking comfort from activities that provide a continuing sense of connection or bond with the person who has died (Klass et al., 1996; Valentine, 2008).

Previous studies of bereavement through murder (Riches and Dawson, 1998) or suicide (Wertheimer, 2001) have identified how stigma, which devalues both the deceased person's memory and the bereaved person's grief (Doka, 1989), may compromise sharing memories and maintaining continuing bonds. Those left behind may have to contend with public narratives, such as those produced by the news media and the inquest, which, having different priorities, do not necessarily reflect the bereaved person's experience (Chapter 2; Guy, 2004; Walter, 2005). For press reporters the priority is to sell newspapers, and the role of the coroner (or the procurator fiscal in Scotland) is limited to identifying the final cause of death. Furthermore, the growing trend towards life-centred funerals, which celebrate the uniqueness of the deceased person's life and character (Cook and Walter, 2005; Garces-Foley and Holcomb, 2005), may pose challenges for those mourning a life cut short, whether by suicide, murder or substance use. With substance use deaths, however, the impact of stigma may be even more compromising, as well as cumulative, in that coping with the stigma of substance use may have already have put the deceased person's and, by implication, the family's reputation into question while the deceased was alive. This situation may be further exacerbated by the co-existence of, for example, mental health issues or criminal activity. Thus, following the stress, strain and uncertainty of living with another's substance use, coping with that person's death is likely to involve painful, even disturbing, memories of their life, as well as the stigma and trauma of how they died (Chapter 1).

This raises the question of how a life and death that give rise to both stigma and painful memories may be remembered and memorialised, both publicly and privately. To consider this question, the chapter examines our interviewees' experiences of memory-making with reference to two main areas that were prominent in their recollections. First, we consider the more public aspects of remembering, particularly the funeral, including how it was organised and what it was like to be there, as well as press reporting of the death and how the inquest (in England) or inquiry (in Scotland) was conducted. The funeral, inquest and press reports all create public constructions of the life and the death. Second, we examine some more private aspects of remembering, including the bereaved person not wanting or finding it too painful to remember, or struggling to salvage fond memories from a mass of bad. We draw on concepts of post-traumatic growth (Joseph, 2012) and continuing bonds (Klass et al., 1996; Valentine, 2008) to identify and discuss how some people actively memorialised the person's identity and life (Valentine and Walter, 2015), both publicly and privately, and incorporated the person's memory into their ongoing life.

Public remembering

The funeral

Studies of funerary and memorialisation trends do not include the experiences of those bereaved through substance use, yet, as this chapter illustrates, their experience can shed light on how available norms and practices provide resources or pose obstacles to mourning. With the growing trend towards life-centred funerals designed to celebrate the deceased person's uniqueness (see, e.g., Cook and Walter, 2005; Caswell, 2011), it might be anticipated that having to organise and take part in a funeral for someone close whose death was the result of their substance use would be a considerable ordeal. For some of those we interviewed, this was indeed the case, due, for example, to feeling overwhelmed and depleted by the circumstances of the death; not wanting to celebrate the person's life; insensitive or stigmatising comments from others, including family members and clergy; family differences of opinion as to how the person died and how openly this should be acknowledged, if at all;

and, particularly for those in treatment or recovery from their own substance use, not being able or deciding not to attend the funeral, or attending but feeling isolated from other mourners. However, for a significant number of interviewees the funeral affirmed the deceased person's identity, providing a formal, public ritual that represented their life as involving far more than substance use. As such the funeral could provide an antidote to stigma, as well as expose the bereaved to public stigma. Drawing on interviewees' experiences of the funeral, both positive and negative, we consider the relationship between the death's link to substance use and i) organising the funeral and ii) taking part in the funeral. Both entailed interacting with other family members and with people outside the family.

Arranging the funeral

Because they had been at the time either in shock or overwhelmed with grief, some interviewees found it hard to remember the funeral.

> *But I don't even remember most of the funeral . . . just wee bits . . . I said to my daughter the next day, did she get a good turn out because I didn't see anybody there? But I think you just go into fast forward.*
>
> (DaughterS)[1]

Others found themselves taking on the task of arranging the event but without really being engaged in what they were doing.

> *I suppose I was going through the motions and so it was like toing and froing to London to do all these things that needed to be done – to arrange the funeral and the wake.*
>
> (FatherE)

Although this may characterise any bereavement, for those bereaved by substance use it was compounded by expecting to encounter stigma, as reported by a bereaved wife.

> *The funeral was incredibly difficult. . . because you've got stigma and shame . . . it's not like going to an elderly person's funeral where you celebrate their life, it's like . . . let's put them in their box and get rid of it quickly and then we can all move on.*
>
> (WifeE)

One mother regretted how her expectation of stigma prevented her from telling her son's friends about his death – the funeral was attended only by family members.

> *But I got this horrible guilt again, but it has worn off a bit, but we didn't tell any of his friends from the past because really I felt they sort of washed their hands of*

him really because of the drugs . . . We just didn't have it in us to go round telling people what had happened.

(MotherE)

For others, in contrast, organising the funeral provided a means of acknowledging and affirming the deceased person's identity. A mother whose daughter's substance use was linked to mental health problems and ultimately led to her suicide told us

[I]t was very very important to us to give her yeah dignity and respect, even though she'd chosen to leave us.

(MotherE)

Another mother used her son's funeral to make others aware of the dangers of drug use.

I think I am too proud to let other people's attitude about drugs affect me and I know at the funeral . . . the guy that was doing the service had said to me . . . what do you want me to say about [son] and I said I want you to say exactly . . . people that come need to know, drugs are an issue they hit my family, they will hit their families or their grandchildren, they have to be aware.

(MotherS)

As noted in Chapter 3, however, being open about the deceased person's drug use could invite insensitive comment. One couple recalled visiting the vicar to prepare for their son's funeral.

We went to his house and he had lost his book with us really, because he turned and said what a selfish thing to do. And I said I don't think it's selfish; I think it's a desperate act.

(ParentsE)

In some cases the bereaved felt at a loss as to how to proceed and needed to rely on the help of others, as one mother reported.

Well both myself and his father were stumped, we had never done anything like this before, didn't really know where to start. So . . . up a wee bit there is a [funeral director], so I phoned up – sorry I met my friend, she was, she kind of took over a lot of things you know.

(MotherS)

However, for one father, the initial sense of relief that came from his son-in-law's offer of help led to unforeseen and uncomfortable consequences.

I think it was just me wanting the whole thing to be over, I didn't want the palaver and I think maybe the stigma . . . I just wanted this to be over and

get everyone away. But [my son's] got a half-sister . . . she married an Ameri-can . . . and he's an elder in the Mormon Church and . . . he said, "I can do a funeral service for you" and I thought . . . well at least he knows him. Let's do that . . . and get it out of the way . . . so we had a service in the Mormon chapel, we chose some hymns and . . . the Mormons took it as an opportunity to sell their religion to my family.

(FatherE)

One mother, wanting to acknowledge and affirm her son's life, felt obliged to step in as she felt that other family members lacked the capacity to do so. She further conveyed how she found the strength to carry out such a painful task through keeping hold of one of her son's possessions.

I had to organise his funeral because . . . it was a mark of respect and a mark of honour to do that for my son and I think I was the only one left capable of actually doing it. My son couldn't do it, his father could not have done it so I did it and I went to the Co-op and I, lovely funeral director, really, really lovely and the only way I got through any of this was, I took my son's toy rabbit from his bedroom and I kept it in my handbag with me and I organised the funeral and hundreds of his friends came.

(MotherE)

For some interviewees personalising the funeral was central to its planning, for example, through demonstrating a particular talent that was important to both bereaved and deceased. One father honoured his son's musical talent.

I had – at the foot of [son's] coffin, I had a stand. I wanted to have a stand that had his guitar there and there were flowers all over it . . . It had his picture at the foot of the coffin . . . and we sorted out some great music – we had some African drumming and it was a very unconventional thing.

(FatherE)

Yet in view of how the person had lived and died, not everyone felt able to go along with the trend of personalising the funeral. One couple recalled their discomfort at being encouraged to do so by the funeral director:

But I think it was kind of hard because they're like don't you want to do something really special? And it's like we don't really know how we feel about all of this, and we need to do something but this is not really a celebration of a life . . . it's never happy people dying, but there's generally a memory of this one amazing person, and it's just like it's not really there.

(ParentsE)

In retrospect, however, a son in another family regretted that the funeral that he, his mother and sister had organised failed to reflect who his father

was and what he would have wanted, in that the family had not felt prepared or up to the task.

*[H]aving to deal with the funeral and all that. Like, we rushed the funeral and . . .
looking back I would have done it a lot different but like . . . we didn't know what
we were doing like, we were chucked in at the deep end, you can't blame us. .*

(SonE)

Yet this son's recollections conveyed how important it could be for the funeral
in some way to reflect the unique person, whether or not this was achieved.
Thus, the trend towards personalised funerals may provide an opportunity for
families to counter stigma through celebrating the deceased's life – despite the
risk of failure or unwanted pressure to personalise.

Taking part in the funeral

Some found the unexpected volume of attendees to be unforgettable and evidence of the value of the deceased's life. Sheer numbers could provide some
comfort.

*[W]hen I got to the crematorium I was absolutely blown away by how many people
were there. There was probably a good 200 people there.*

(MotherE)

*I didn't really know that [son] had as many friends. The chapel was stowed . . . And
I've never seen it so stowed which is a testament to . . . him as a person that as
many people attended his funeral which as I said I didn't know that he had that
many people in his life . . . At the time you don't appreciate it but it was kind of a
wee comfort.*

(MotherS)

For one mother such a large turnout enhanced the funeral's role as way to say
goodbye.

*We had hundreds and hundreds of people there, it was just gobsmacking . . . there
were so many people they couldn't fit them in the chapel, they had to open the doors
and just let people listen in from outside . . . It was a beautiful, beautiful service, we
had a wicker coffin, and I was able to say goodbye to him properly, and his family
as well, we all did.*

(MotherE)

However, for one daughter, the experience of a "packed" funeral raised the
painful gap between her own experience of living with her mother's alcohol
use and those who had known her mother in happier circumstances.

[T]he funeral was packed . . . I was sitting there thinking why do all these people like her after the type of mother she was you know? Why was she nice to all these people when she wasn't like that with us? And I just thought to myself . . . when I was sitting there and I was crying . . . I still tried to think of the in-betweens, how she got from there to being the way she ended up.

(DaughterS)

For some, the funeral was valuable in revealing a side of the person's character that they had not previously known. One mother reported how she and her husband learned more about the positive aspects of their son's life from his friends' contributions to the service.

We didn't know what music he liked . . . his best friend said he loved to dance, and we didn't know he liked dancing. We never saw that side of him at all. We always got the – not grumpy but miserable-looking face, drugged [son], you know? His friends said he was the life and soul. He was a great laugh . . . and he was very caring and I never saw that in him really.

(MotherE)

For one brother, his sister's funeral was particularly important in demonstrating to other family members who had shunned her just how much others in her life had valued her.

[I]t was a really eye-opener for the rest of the family actually at the funeral because the brothers had written her off as this, she's a drunk or, can't remembers seeing her blah blah, they are both academics, they both are head people and not heart people and this is her family . . . I thought where have you been for the last 18 months?

(BrotherE)

Another brother conveyed how positive memories of the deceased could be a welcome distraction from the way the person died.

[S]ome nice stories that were told and that made a big difference because you were remembering how good a guy he was so you weren't thinking about the tragic circumstances that led to him dying.

(BrotherS)

However, the emphasis on celebrating a life could be bittersweet. One father recalled how the many fond memories expressed at his son's funeral paradoxically brought to mind how his son had been unable to appreciate and value himself.

My daughters said something and his mother wrote a little history of his life, which someone else read out, and parts of it were funny . . . and the music was very celebratory – and the wake in the evening was just all light music, so the evening was a celebration too. And so many people were there who valued him – but what is sad is that he didn't seem to be able to accept that.

(FatherE)

For others, recollections of the funeral emphasised family tensions over how the person died. For example, one daughter recalled with some bitterness how her father's family, while expressing grief at the funeral, had distanced themselves while her father was still alive.

My mum's side of the family were all Catholic, my father wasn't religious and so we decided not to have a Catholic funeral, so we had a normal Church of England funeral, very full, there were lots and lots of people. And my dad's family were very upset, but I still felt very bitter towards them. And then we just had a normal wake and then everyone just disappeared once again.

(DaughterE)

Another daughter conveyed similar feelings about relatives who had avoided her mother while she was alive.

I didn't enjoy afterwards about relatives came round because they'd never bothered with us and probably for the same reason, because they knew mum was drinking . . . It just felt very false and I felt very impatient about it and I thought "let's just get this over with".

(DaughterE)

A sister recalled feeling angry at her mother for not disclosing the cause of her brother's death, while acknowledging potential for stigma. Relatives who attended the funeral were under the impression that her brother had died of a heart attack.

Not so sure that some of the aunties or uncles really knew why they were there or why he'd died. But then that was my mother keeping it from them. I think a lot of people thought he he'd had a heart attack . . . I did feel anger sometimes towards that, because I think "why didn't she just tell them he was an addict?" . . . I think because, I don't know, there's still a stigma there I think.

(SisterE)

Similarly, an ex-wife recalled how her brother-in-law imposed silence about the real cause of her husband's death.

[A]t the funeral, I wanted any proceeds to go to NACOA.[2] I didn't do work for them at the time but I wanted, my brother-in-law didn't want the funeral to have any mention of alcohol. So I thought, "this is what killed him but we're not to mention it".

(Ex-wifeE)

However, where the cause of death was no secret, family members were vulnerable to insensitive comments.

> *I was really upset when we got to the funeral . . . my sister heard somebody say, "I don't know why they are so upset", which to me to this day, one, I don't know who said it and secondly what a thing to say and in ear shot.*
>
> (DaughterE)

Some interviewees felt unable to attend the funeral for fear of encountering family members, with whom they had issues.

> *I didn't go to the funeral and I had good reason to, when my partner was alive they [family]caused her nothing but grief . . . and I didn't want any arguments and I know from past experiences there would be some.*
>
> (PartnerE)

> *I didn't go because there was no way I wanted to see my family. It's a bit of a regret but I was scared stiff of there being like an Eastender [a popular British TV soap], so what we did instead was my husband and I went to the Cathedral and that's still somewhere and it's a bit of a sanctuary for me now, I work right next door.*
>
> (SisterE)

One interviewee chose not to go to her ex-husband's funeral because she sensed that his family were wondering if she was in some way to blame for having ended the marriage.

> *I did get a distinct feeling from speaking to his son that possibly . . . was it my fault that I'd split up with him, I should have supported him more? I just thought if I went along there I was going to be, I don't know just, like a needle maybe.*
>
> (Ex-wifeS)

Interviewees who were former substance users and who chose to attend the funeral could feel as though they did not belong – neither to the family nor any longer to the drug-using community.

> *When I went to the funeral . . . I saw lots of people I used to use drugs with, and . . . I felt like I was in the wrong place. I had all these feelings of I wasn't an actual part of the family . . . So I had to keep quiet, out of the way, which felt really odd because my friend and I were really close.*
>
> (FriendE)

Another friend, as well not feeling part of the family, felt that his closeness to the deceased was unacknowledged.

I had all these feelings of I wasn't an actual part of the family, but I was really, really close to [deceased friend] and that just felt really odd. It felt like . . . [deceased brother's] jealousy had got in the way of something [we] shared that may have been, well, from my point of view I would rather have had a different song being played, but I wasn't a member of the family so I couldn't say anything. So I had to keep quiet, out of the way.

(FriendE)

Seating is a key way family and non-family are demarcated and displayed at British funerals (Walter and Baily, in press), but this social ordering could be distressingly disrupted. A brother who was in prison for a drug offence at the time of her brother's death, although allowed to attend the funeral, was separated from her family.

I didn't really grieve at all, because I was handcuffed to a prison officer and your family sits this side and friends sit this side. I was next to a prison officer, so I chose to sit over the other side in the front row, but the other side, because the prison officer isn't really my family.

(Brother E)

One mother, although remembering little of the funeral itself, was still struggling with her outrage at her oldest son having been handcuffed to a prison officer and therefore separated from the family.

[A]nd do you know my oldest son was handcuffed . . . in the [cemetery] for out stealing for drugs. He didn't have a gun, he didn't have a knife, he never murdered anybody and they had him handcuffed, standing watching his brother going down the . . . So cold and awful.

(MotherS)

Thus, for those who are mourning a substance use death, our study found that in many, although not all, cases the personalised funeral as a celebration of the life could to some extent counter the stigma these deaths attract. The funeral validated both their loss and the person's memory; it could also reveal aspects of the person's life that the family had not been aware of. However, personalised funerals did not always feel appropriate in light of how the person's substance use had affected family relationships. The funeral could also exacerbate disagreement within the family whether to acknowledge the cause of death. Anticipating family tensions, some mourners stayed away. Attending the funeral could also be an isolating experience; some former substance users felt separate from the family, whereas some who *were* family members were separated through being handcuffed to a prison officer.

The press

Whereas in some cases the funeral helped to counter stigma, press reporting (as well as the inquest and police investigations) could reinforce it. Previous research has found that in reporting cases of murder or manslaughter, the deceased's behaviour may be implicated or questioned in some way, thus 'spoiling' the identities of both the deceased and, by association, their family or other intimates left behind. The implication is that they, too, must be at fault in some way, a message likely to compound grief with feelings of failure, self-blame and shame (Goffman, 1963; Riches and Dawson, 1998). Where substance use is implicated in the death, it is even more likely that what is perceived as the deceased's self-destructive behaviour, which may have involved illicit activity, will be subject to social censure, implicating family or close others. Furthermore, these families may already have blaming themselves for failing to prevent the person's substance use while they were alive; press reporting of the death that demonises drug users and their self-destructive behaviour (Grace, 2012; Guy, 2004, p. 52), can exacerbate these feelings.

Instead of encouraging the reader to identify with the grieving families, this kind of press reporting tends to distance the reader from both the deceased's self-destructive and/or illicit behaviour and the family that failed to prevent it. Thus any attempt to find coherence and meaning in the death may be inhibited by press reporting that reinforces and promotes popular myths and negative stereotyping of substance use. Yet the images and assumptions attached to those who use substances take little, if any, account of the person behind the image or of the families who were close to them. Rather these stereotypes tend to set both the deceased and the bereaved family apart from the rest of the community. Thus, in spite of media attention potentially providing a means of sharing the family's story with the wider community, the language used to convey these stories may have the effect of isolating the family from the wider community, subjecting them to a profoundly lonely grieving experience, with particularly negative consequences for grief, health and well-being (Guy and Holloway, 2007).

Such reporting may in part reflect press working practices in which reporters may draw on particular formulas to enable them, within the constraints of time and column space, to come up with an attention-grabbing story. As one journalist explained during a focus group, "[Y]ou've got three minutes to do that story before you go on to another 'suspicious' death". However, for the bereaved people left behind, such attention-grabbing phrases may effectively depersonalise the deceased, someone they knew and loved as being far more than their substance use.

> *'Drug addict dies', you know, 'overdose.' Just nothing about the person, just a headline.*
>
> (MotherE)

> *I just read, 'Unemployed man dies of drug overdose' and read down through and it was [my son] . . . so I wrote to the editor of the paper, I explained the type of*

person [my son] was and I don't think the main point about him was that he was unemployed, there was more to [him] than an unemployed man . . . but the type of journalism nowadays . . . they don't want this person to be an ordinary nice person, they want him to be a drug addict.

(FatherE)

One focus group participant, a practitioner from Scotland, drew attention to the lack of control that family members may have over press coverage.

They will ignore the immediate family and go around some of the pals . . . to try and get a more sensational story . . . and the family won't be contacted but will read about it in the press . . . and they might phone up and complain . . . and the editor will say. . . "we have sources", and again that feeling of powerlessness . . . that somebody's name is being dragged through the mud.

[PractitionerS]

Indeed one married couple reported having been hounded by the local press for a story about their son's drug-related death.

They were so intrusive – they approached us . . . We said, "Absolutely not. We're not talking about it at all. We don't want any photographs; we don't want anything" . . . They visited all our neighbours for about three doors either side and asked them for comments.

(ParentsE)

However, it would seem that some journalists may be willing to take a more collaborative approach, as in the case of a father who recalled a more positive experience of a journalist consulting him in advance and giving him the opportunity to change the wording.

She [newspaper reporter] said, "Could we work together on this as opposed to me just writing the story up and disappearing back to the office?" She wrote it up and . . . emailed it to me and . . . said, "Have a look. Is this okay? This will be the actual wording . . . Is there anything you want to change?" So I changed a few things.

(FatherE)

In another case, the local press were willing to publish a bereaved mother's report of her son's death.

It was very generous. But you see we're a bit – a friend who works for the media had said to us what you need to do is write out whatever, a report for the reporters, and then give it to them. And then it's up to them if they use it or not.

(MotherE)

Thus, although the press has a duty to report unexplained deaths, how those deaths are reported affects the public's attitude, not only to the deaths, but to those left behind. If carried out sensitively, such reporting could potentially be helpful to those left behind, particularly in countering stereotypes by providing a more nuanced and humane version of events. In any case, press reporting is likely to contribute to how the bereaved person remembers the death in a more personal sense. Although most research on meaning-making in bereavement sees this as an individual (Neimeyer and Sands, 2011) or, at most, a family (Nadeau, 1998) process, our research confirms Walter's argument (2005) that public – and hence authoritative – accounts of the death given at the funeral, at the inquest and in press reports can be very significant for individual family members. These public accounts can legitimate or undermine personal constructions of meaning, causing family members considerable relief or, alternatively, distress, anger or shame. Especially after deaths that interest the press and/or officers of the state, meanings cannot be constructed by mourners in isolation.

The following section illustrates and discusses the implications for the more personal dimension of remembering a life that has involved substance use and which also tends to attract a bad press, both literally and metaphorically, in terms of social stigma.

2. Personal ways of remembering

Problems with remembering

Bad deaths (Seale and Vandergeest, 2004) more generally, for example, premature, sudden, violent and unexpected deaths, are likely to pose difficulties for memory-making and regaining a sense of control and continuity in life (Valentine et al., 2016). However, where substance use has been involved in the death, as discussed in the previous sections, piecing together and preserving memories of the deceased person's life may run into the spoiling of these memories due to bad press and social stigma. The experience of stigma may also inhibit sharing one's grief with sympathetic others in a way that provides understanding and validation of one's loss (Rosenblatt, 1993). In addition, regaining a sense of control and continuity in life may be compounded by the relationship with, and the lifestyle of, the user prior to their death. Indeed, for many bereaved families the effects of having lived with the person's substance use (Orford et al., 2010), including coping with the possibility that it may eventually lead to their death (Oreo and Ozgul, 2007), can be profoundly destabilising. Thus, regardless of stigma, painful memories related to the deceased person's life may also make it difficult to take support and comfort from fond memories so as to achieve some sense of 'normality'.

For example, one father conveyed how remembering could evoke the sense of waste of life, lost hope and feelings of bitterness, whereas for a daughter, her mother's death put an end to the hope that things would change.

And having to get to grips with a death was overlaid with having to get to grips with the fact that this sense of a waste . . . obviously there's always the grief associated with the death of anybody – and any close member of your family, it's entirely normal but also a sense of total waste of a life.

(FatherE)

I always had this hope. . . [but] when she died I lost my hope and that was the saddest thing for me that there was no hope of it ever getting better.

(DaughterE)

Another daughter related her struggle with difficult childhood memories of her mother's drinking and how these have left her with ongoing feelings of bitterness.

I am still bitter about my mother, very, very bitter. And still to this day blame her sometimes with memories that I have from my childhood. And even . . . if she was alive today in this room honestly I wouldn't even be sitting in the same room as her, mother or not.

(DaughterE)

For one daughter, such difficult memories were tinged with sadness in that they were a painful reminder of the mother daughter relationship she had never experienced.

[M]y mum is my mum but I have a lot of very, very disturbing nasty memories of her . . . I wouldn't say I have got a lot of good memories of my mum at all. It's sad . . . because . . . I would die to have a natural friendship, a mother and daughter bond you know the way a mother should have with her daughter.

(DaughterS)

Yet bad memories could to some extent be tempered by remembering and holding on to positive images of the person, whether prior to their substance use, or positive aspects of their character in spite of their substance use.

Yes I do believe that she's like in heaven having a lovely time, she's fine, she's happy, she's peaceful . . . Do I imagine her with a drink or without a drink, or would I rather have her back with all her problems? . . . you wish them to all go away – but then you just think well I'd like to have her back with that just as much as without.

(DaughterE)

[H]e was a kind hearted kid at heart you know he loved bacon, he loved making soup, just lots of things. Very good to his grandmas, he just adored his grandmas and he had a lot of kindness in him that way so yeah. But he had a lot of badness as well. You know I can't rule that out you know. But I mean he paid for it.

(MotherS)

In the absence of any positive memories of one's own it was possible to rely on those of someone else, as conveyed by a daughter in relation to her mother's memories of her father.

> *[A]nd I don't like the fact that my memories of him are all bad probably. I do remember seeing him drunk a lot the time. I have seen him be abusive, and just not particularly nice which I don't think is a fair or a true reflection of what he was actually like when he was sober . . . like I do believe Mum when she says that there was a completely different side of him.*
>
> (DaughterS)

In some cases it was a matter of time before the bereaved person was able to remember and take some comfort in positive aspects of the person, for example, as a result of sorting through photographs of the deceased and selecting some to retain, or through realising the value of remembering the good rather than dwelling on the bad

> *And I got the pictures down and I just started going through them and I just thought, "oh that would be nice, I'll put that in there". I've not even got a picture of my brother in my house at all. Not until yesterday.*
>
> (SisterE)

> *[F]or a long time it was really hard to remember that there was any good times because it was just so awful before he died . . . But I can remember now . . . we had holidays and . . . he was quite charismatic . . . if you walked in to a room . . . you would notice him because he was quite sort of bright and bubbly and . . . people did like him . . . So there are things . . . that were nice and it's important to remember that instead of . . . dwelling on all the awful times.*
>
> (MotherE)

Memories of bad experiences could be made good through trying to understand what drove the person to drink, develop empathy for their predicament and discover more about the family dynamic and its influence on one's own life and identity.

> *But for me I wanted to understand why she drank I think. And I don't have the answer to that but I think I've tried to understand a bit more about my family dynamic and why things are as they are, tried to piece together some of the things that perhaps weren't quite as they seemed . . . and learning . . . I suppose going by how it's affected me in positive ways as well as negative ways . . . would I change it? I would like some of the pain to have gone away but I do think the relationship I have with my mum has made me who I am and from that I have a lot of empathy*

and . . . determination and it has enabled me to achieve things in my life that I may not have done otherwise.

(DaughterE)

Painful and disturbing memories could inhibit finding any comfort or support in fond memories, leaving the bereaved person to cope with wanting to forget, yet being unable to, and so being at the mercy of intrusive memories. Yet in spite of painful memories of the person's life and the manner of their death, in many cases, and particularly with time, the bereaved people we interviewed were also able to remember and cherish more positive aspects of the person and their relationships with them. As Rosenblatt's (1983) reminder theory indicates, bereaved people need to strike a balance that works for them between remembering and forgetting. The following section explores this process in more detail in terms of living with both the good and the bad through actively including the deceased person in their ongoing lives or continuing the bond with the person.

Remembering and continuing bonds

Numerous research studies have drawn attention to the diverse and creative ways in which bereaved people may remember and include deceased loved ones in their ongoing lives (Klass et al., 1996; Valentine, 2008). Although physically absent, those who have died may remain socially present in the lives of the living via culture-specific forms of memory-making. In contemporary society, these may include memory boxes, grave displays, internet memorial sites and social media networks (Walter et al., 2011); or more traditional forms, such as giving a eulogy, tending the grave, displaying photographs of the deceased and retaining keepsakes (Hallam and Hockey, 2000; Gibson, 2008; Valentine, 2008). Memory-making and ongoing relations can be internal and/or private, as well as co-created or shared (Unruh, 1983; Klass et al., 1996; Walter, 1996).

With substance use deaths, however, as conveyed in the previous section, such sharing may be inhibited by painful memories of the deceased person's life and death, compounded by the social stigma of substance use. Indeed, from a psychological perspective, substance use bereavement has all the hallmarks of traumatic loss; the mental and emotional and physical toll it may entail risks complicated grief disorder developing (Stroebe and Schut, 2005–06). As well as identifying people's vulnerability, however, general bereavement research has identified resilience (Bonanno, 2009) and post-traumatic growth (PTG) processes by which adversity can lead to increased psychological well-being (Joseph, 2012; Calhoun and Tedeschi, 2006).[3] This perspective reveals a mixed and complex picture of the way people grieve, in which vulnerability and resilience may go hand in hand; responses to traumatic loss can be both diverse and creative (Valentine and Walter, 2015).

The following examples confirm diversity and creativity to be well represented in our sample, challenging any type casting of substance use bereavement and highlighting the capacity of individuals to subvert negative cultural stereotypes. Indeed several of those we interviewed talked about how they had actively challenged stereotypical assumptions about substance use. Examples of this included learning about, working with and setting up organisations to support those living with and/or bereaved by addiction; writing/publishing autobiographies and personal testaments of the deceased person; and, more privately, through tending the grave, talking, writing letters and poems to the person and creating memory boxes.

One daughter whose father turned to alcohol after losing his job took the opportunity to study alcoholism as part of her psychology studies in order to better understand her father's suffering. When her father died, she had qualified as a teacher and felt she was in a position to do more with her understanding of alcoholism,

> [M]aking people more aware and . . . not demonising somebody who has an addiction and trying to help them . . . What I'm hoping to do kind of going on from the fact that lots of adults find it quite uncomfortable to talk about alcoholism and death through alcoholism I am hoping . . . to start building educational materials for PSHE [Personal, Social and Health Education] so that teachers can kind of pull it off of the database or system . . . A good set of resources with backed up material, so they can feel more confident in teaching about alcoholism.
>
> (DaughterE)

A mother whose son died of a drug overdose used her writing skills to challenge negative stereotypes. She produced articles and a book about her experience of bereavement. She wanted to dispel misconceptions about drug users as well as encourage others to share similar experiences rather than be silenced by shame.

> I'm not so much interested in telling the story; I'm more interested in telling the story so that people who might have been – who might be like me and might have preconceptions about people who use drugs and the families of people who use drugs . . . and the stigma and the shame and all of that. By telling the story it's, you know, and then somebody else tells their story. And before you know it, it's no longer such an area for shame that people can be more open. I don't like doing it. It's just not comfortable with me. But it's only by doing that that you can say, and, you know, I'm proud of my son.
>
> (MotherE)

Although uncomfortable and challenging, bearing witness to her son having achieved far more in his relatively short life than his drug use might suggest enabled her to remain connected to him and to take pride in his memory.

A man whose brother died of an alcohol and drug overdose was hurt and angered by stigmatising responses which devalued the life of his brother and others like him, but he also believed such responses to be rooted in lack of awareness. Before his brother died, he had begun a degree in media studies, and following the death he decided to use his studies to explore substance misuse.

> *[A]nd it was going into the second year that I, kind of, started to use film and when my brother died, I was just going into my third year which is all self-motivated work . . . and that's why I felt I was in a position at that point to do that, because that was the only thing on my mind.*
>
> (BrotherS)

So he used his film-making skills to portray the complex human story behind the stereotypes.

> *[T]he fact that you could say, well, it's better that they're dead than have somebody that uses in society and that . . . I think that really affected me to the point where that's what really made me want to explore drug use further and understand it better and help . . . raise awareness of the fact that, you know, he was a person. So . . . I was doing a lot of work on drug-related deaths and trying to find out as much about that, speaking to police and practitioners and people that are still using and as many people as I could, using film, to try to gather all these stories together and make this film, sort of tackle . . . those kind of stereotypes and ignorance and stuff in society.*
>
> (BrotherS)

A child of alcoholic parents who were leading particularly chaotic lives relied on her uncle for her support, to whom she felt very close. Yet after her uncle's death, her grief was not deemed to warrant any kind of bereavement support.

> *[J]ust because you're not a partner or a child, it doesn't mean that you're not going to feel it just as much.*
>
> (NieceE)

Yet rather than remaining isolated, her experience of being alone with such painful and distressing feelings drew her to the plight of others in similar circumstances, prompting her to train and work as an online mentor with an organisation for supporting children of alcoholic parents.

> *So, you know, there's lots and lots of children and young people going through that kind of thing, so I just really want to help them and be there so they can speak out, so they're not trapped and alone. And maybe speaking to somebody on [helpline] or speaking to me on [helpline] will give them the confidence to change more about their life and maybe get more help and get out of the situation. So that's why I really*

want to do it. And for me that just makes me feel better knowing I can have that impact in somebody's life.

(NieceE)

By making a positive contribution to those, like herself, whose need for support tended to be overlooked, she was able to recover a sense of personal effectiveness.

Although resources such as sharing with sympathetic others (Doka, 2002) may be unavailable, these examples have demonstrated how meaning may be found through using personal and private grief in the service of a wider public and social agenda. In drawing on and developing existing skills, these interviewees were able to recover and repair an important thread of continuity in their lives, which included their own and the deceased's identity, as well as their relationship with them.

For others, however, meaning and continuity were recovered through more private activities that enabled them to develop a positive and comforting connection to the deceased. One mother whose son's substance use was linked to a psychological disorder was unable to find support either before or after his death. Thus, her experience of both his life and death was particularly undermining and destabilising. Yet in spite of considerable self-recrimination and an absence of fond memories, she found her own way of recovering her connection to both her son and her own sense of integrity.

Well, it started with our first Christmas without him. I decided I would write him a Christmas card . . . and I dated it and sealed it and put it in a tin. And then just after Christmas with what would have been his 24th birthday, I sent him a birthday card and put it in the tin. And then when it got to . . . the first anniversary, I wrote him a letter. And in the letter I told him about things that had happened since he died . . . So this tin is now absolutely full of letters that I've written to [him] and I'm still writing to him. And that's the only way that I can kind of hold it together.

(MotherE)

Another mother was motivated to record and store her memories of her son's life for fear that she might forget him.

Sometimes you force yourself to do it because you are frightened you will forget . . . another thing that panics me is I when I think to the future, of when I am older and he's been dead like twenty years, thirty years, forty years . . . as well thinking how long he won't have been here for . . . so, I write things down . . . I wrote a whole list of everything he liked . . . favourite numbers, colours . . . I wrote it all down for him and put it away with all his personal things. I've got them in boxes in the loft.

(MotherS)

The niece grieving her uncle's death, although finding support through supporting others, also pursued private ways to continue her relationship with her uncle. Well documented in the reported experiences of bereaved people more generally (Hawkins, 1999; Francis et al., 2005; Valentine, 2008), these included writing songs, poems and letters to him and depositing these at his graveside.

> *I write a lot of poetry about him. I write a lot of songs about him . . . I just go on my own and have a chat to him. You know it's weird having a one-way conversation. That was the biggest thing I had to learn to get over when I first start visiting his grave was that he's not talking back, but in a way I feel like he is anyway . . . because he understood me so well that I can almost imagine the kind of reply he would give me anyway. So it's definitely really important to just keep that going, like I will always visit his grave. If it's looking tatty, I would always clean it. I will always take him a little something and I will write him a poem or a letter, because he's still there.*
> (NieceE)

In actively continuing their relationship with the deceased person, these interviewees found reinforcement following an experience that could potentially undermine the integrity of both bereaved and deceased, as well as the relationship between them.

Conclusion

This chapter posed the question of how a life and death that involved substance use may be remembered and memorialised, both publically and privately, given both the cultural assumptions and related stigma of a substance use lifestyle and the disturbing memories of the person's life and death. The examples we have cited illustrate diversity and ingenuity in how those bereaved by substance use may remember and memorialise, how they may struggle to forget as well as to remember, and how they salvage the good memories from the bad and retain a balanced picture of the person. For some, memory-making was a solitary experience, whereas for others it relied on other peoples' memories and support. In all cases it involved a complex process that evolved over time and was shaped by the circumstances of the life and the death and how far these left the bereaved person with painful and disturbing as well as fond memories. Memory-making was also affected by the funeral and by press reporting, whether directly through the death being reported in the local press or indirectly through the way such deaths are reported more generally. Beyond these more public accounts, however, memories were largely grappled with, endured, avoided, shaped and stored in private, although they could motivate projects designed to raise awareness, challenge social stigma and support others in similar situations. In the process a significant number of those we interviewed conveyed how, in spite of the obstacles to remembering a substance use death, they had been able to find a place for the deceased person in their ongoing lives.

Notes

1   'E' and 'S' indicate interview participants in England and Scotland, respectively.
2   National Association for Children of Alcoholics

3   Beyond the scope of this chapter, Calhoun and Tedeschi (2006) have identified five dimensions of psychological growth, including personal strength, new possibilities, improved relationships, spiritual change and appreciation of life.

References

Berger, P. L., and Luckman, T. (1967). *The Social Construction of Reality*. London: Allen Lane.

Bonanno, G. (2009). *The Other Side of Sadness*. New York: Basic Books.

Braun, M. J., and Berg, D. H. (1994). Meaning reconstruction in the experience of parental bereavement. *Death Studies*, 18, 105–129.

Calhoun, L. G., and Tedeschi, R. G. (Eds.) (2006). *The Handbook of Posttraumatic Growth: Research and Practice*. Mahwah, NJ: Lawrence Erlbaum Associates Publishers.

Caswell, G. (2011). Personalisation in Scottish funerals: Individualised ritual or relational process? *Mortality*, 16(3), 242–258.

Cook, G., and Walter, T. (2005). Rewritten rites: Language and social relations in traditional and contemporary funerals. *Discourse and Society*, 16(3), 365–391.

Doka, K. (2002). How we die: Stigmatized death and disenfranchised grief. In Doka, K. (Ed.). *Disenfranchised Grief: New Directions, Challenges and Strategies for Practice*. Champaign, IL: Research Press, pp. 323–336.

Doka, K. (Ed) (1989). *Disenfranchised Grief: Recognising Hidden Sorrow*. Lexington, MA: Lexington Books.

Garces-Foley, K., and Holcomb, J. S. (2005). Contemporary American funerals: Personalizing tradition. In Garces-Foley, K. (Ed.). *Death and Religion in a Changing World* (pp. 207–228). New York, London: M.E. Sharpe.

Gibson, M. (2008). *Objects of the Dead: Mourning and Memory in Everyday Life*. Carlton, Victoria: Melbourne University Press.

Goffman, E. (1963). *Stigma: Notes on the Management of Spoiled Identity*. London: Penguin.

Grace, P. (2012). *On Track or Off the Rails? A Phenomenological Study of Children's Experiences of Dealing With Parental Bereavement Through Substance Misuse*. Unpublished PhD Thesis. University of Manchester, UK.

Gunaratnam, Y., and Oliviere, D. (2009). *Narrative and Stories in Health Care: Illness, Dying and Bereavement*. Oxford: Oxford University Press.

Guy, P. (2004). Bereavement through drug use: Messages from research. *Practice*, 16(1), 43–54.

Guy, P., and Holloway, M. (2007). Drug-related deaths and the 'Special Deaths' of late modernity. *Sociology*, 41(1), 83–96.

Hunsaker Hawkins, A. (1999). *Reconstructing Illness: Studies in Pathography*. West Lafayette, IN: Perdue Research Foundation.

Joseph, S. (2012). *What Doesn't Kill Us: The New Psychology of Posttraumatic Growth*. London: Piatkus.

Klass, D., Silverman, P., and Nickman, S. (Eds) (1996). *Continuing Bonds: New Understandings of Grief*. Washington, DC: Taylor and Francis.

Neimeyer, R. A. (Ed.) (2001). *Meaning Reconstruction & the Experience of Loss*. Washington, DC, London: American Psychological Association.

Neimeyer, R. A., and Sands, D. (2011). Meaning reconstruction in bereavement. In Neimeyer, R. A., Harris, D. L., Winokuer, H. R., and Thornton, G. F. (Eds.). *Grief and Bereavement in Contemporary Society: Bridging Research and Practice*. New York: Routledge, pp. 9–22.

Oreo, A., and Ozgul, S. (2007). Grief experiences of parents coping with an adult child with problem substance use. *Addiction Research and Theory*, 15(1), 71–83.

Orford, J., Copello, A., Velleman, R., and Templeton, L. (2010). Family members affected by a close relative's addiction: The stress-strain-coping-support model. *Drugs: Education, Prevention and Policy*, 17(1), 36–43.

Parkes, C. (1988). Bereavement as a psycho-social transition: Processes of adaption to change. *Journal of Social Issues*, 44(3), 53–65.

Riches, G., and Dawson, P. (1998). Spoiled memories: Problems of grief resolution in families bereaved through murder. *Mortality*, 3(2), 143–160.

Rosenblatt, P. (1993). Grief: The social context of private feelings. In Stroebe, M., Stroebe, W., and Hansson, R. (Eds.). *Handbook of Bereavement Research: Theory, Research and Intervention*. Cambridge: Cambridge University Press, Ch. 7, pp. 102–111.

Seale, C. (2000). Changing patterns of death and dying. *Social Science & Medicine*, 51, 917–930.

Seale, C., and Van der Geest, S. (2004). Good and bad death: Introduction. *Social Science & Medicine*, 58, 883–885.

Stroebe, M. and Schut, H. (2005–6) Complicated Grief: a conceptual analysis of the field. Omega 52(1), 53–70

Unruh, D. R. (1983). Death and personal history: Strategies of identity preservation. *Social Problems*, 30(3), 340–351.

Valentine, C. A. (2008). *Bereavement Narratives: Continuing Bonds in the Twenty First Century*. London, New York: Routledge.

Valentine, C. A., Bauld, L., and Walter, J. A. (2016). Bereavement following substance misuse: A disenfranchised grief. *Omega: The Journal of Death and Dying*, 72(2), 283–301.

Valentine, C., and Walter, J. (2015). Creative responses to a drug- or alcohol-related death: A socio-cultural analysis. *Illness, Crisis, & Loss*, 23(4), 310–322.

Walter, T. (2005). Mediator deathwork. *Death Studies*, 29(5), 383–412.

Walter, T. (1996). A new model of grief: Bereavement and biography. *Mortality*, 1(1), 7–25.

Walter, T., Hourizi, R., Moncur, W., and Pitsillides, S. (2011). Does the internet change how we die and mourn? Overview and analysis. *Omega: Journal of Death & Dying*, 64(4), 275–302.

Wertheimer, A. (2001). *A Special Scar: The Experiences of People Bereaved by Suicide*. London: Brunner-Routledge.

Chapter 5

The diversity of bereavement through substance use

Lorna Templeton, Jennifer McKell, Richard Velleman and Gordon Hay

Introduction

We have already seen (Chapter 1) that the experience of being affected by another's substance use is a variform universal, meaning that there are both commonalities to the experience (the 'universal' part), but also differences (the 'variform' part) that relate to a range of factors. Research has focused on culture, faith and religion (e.g., Ahuja et al., 2003; Orford et al., 2015); gender (e.g., Scaife, 2008; Velleman et al., 2007) and a range of relationships between family members and substance using others, including parents (Velleman et al., 1993), partners (e.g., Velleman et al., 1993; Velleman et al., 2007), children (e.g., Adamson and Templeton, 2012; Barnard and McKeganey, 2004) and grandparents (e.g., Barnard, 2007; Templeton, 2012). Chapter 1 further confirmed the presence of this variform universal in our bereavement through substance use sample and indicated how such ideas could be applied to experiences after death. This chapter will consider in more depth how different kinds of interviewees differed in what they said about the impact their relative's or friend's death had on them.

In so doing, this chapter builds on other research which has considered diversity in bereavement and how this could be applied to our bereaved through substance use sample. For example, one U.S. study with 947 adults comprising immediate family, extended family and friends bereaved following a death from homicide, accident or natural causes found that separation distress was greater than traumatic distress for all three groups and that traumatic distress was highest for immediate family followed by friends and then extended family (Holland and Neimeyer, 2011). In a UK study, Wertheimer (2001) interviewed 50 adults bereaved by suicide. The parents, siblings, children and spouses in her sample of 36 women and 14 men allowed Wertheimer to reflect on the particular issues raised in bereavement for each of these four groups; we will return to these in more detail later in the chapter. More recently, an Australian study investigated the potential for post-traumatic growth (defined as "positive psychological change experienced as a result of the struggle with highly challenging life circumstances" [Tedeschi and Calhoun, 2004, p. 1]) in 146 bereaved adults, comparing those who had a lost a first-degree relative, a second-degree relative or a friend (Armstrong and Shakespeare-Finch, 2011). Their findings indicated that post-traumatic growth was influenced by both the severity of the trauma associated with the death and the relationship of the bereaved to the deceased. Whereas the latter was a weaker association, overall growth appeared to be greater for first-degree relatives (who also reported the greatest levels of trauma), followed by friends and then second-degree relatives.

To consider how such findings might apply to adults bereaved through substance use, we explore next some characteristics of our sample, identifying aspects of interviewee experiences which appear to be more prominent in each sub-group, or which appear to differ from the other sub-groups. Inevitably there is diversity within each sub-group; for example, how close interviewees were to the deceased, their understanding of the person's substance use, the nature of the substance use, the impact of the substance use on both interviewee and deceased, when the death occurred and how traumatic it was. We also consider some differences in the profile of each group. Finally, the aim of the bereavement through substance use study was not to interview a representative sample of adults bereaved in this way, so we cannot claim that the findings presented here are representative either. Nevertheless, our data do confirm the idea of there being a variform universal, an idea we believe is worthy of further investigation and should be valuable when developing help and support for a group of bereaved people who have been largely isolated and marginalised.

Relationship between bereaved and deceased

Interviewees were related to the person who died in six separate ways (Table 5.1). Variation across these six groups may contribute to differences in their experience of bereavement: for example, friends were more likely to be male; children and extended family members were more likely to talk about an alcohol-related death and for that death to involve a female, whereas for other groups deaths were commonly associated with drugs or were more equally distributed between alcohol or drugs; only one parent was in treatment or recovery. On the other hand, with the exception of deceased parents or extended family members, the deceased were more likely to be male in this regard mirroring official data for the UK.

Parents

> *It's against the law of nature isn't it? Parents shouldn't bury their children.*
>
> (ParentsE)

It is generally recognised that in contemporary Western societies, the loss of a child is the most difficult bereavement (e.g., Riches and Dawson, 1996, 2000;

Table 5.1 Characteristics of groups within the sample1

Group	Location of interviewees	Interviewee characteristics	Characteristics of the deceased	When the death occurred	Cause of death
Parents (N = 56) Includes 6 couples	30 England, 26 Scotland	42 female, 11 male Age at interview: range 46–75; mean 61 Age at death: 40–68; mean 53 1 father in treatment	Gender: 44 male, 5 female Age: 16–49; mean 29	Approx. half of deaths 2000–2009; one-third 2010-onwards; rest in the 1990s	Approx. half the deaths overdoses (usually illegal drugs); 8 suicides (5 hanging); 8 involved alcohol; 2 murders; rest were various
Children (N = 21) Includes 1 adopted child, and 2 siblings of the same parent	19 England, 2 Scotland	16 female; 5 male Age at interview: range 22–58; mean 37 Age at death: 9–55; mean 27 4 in treatment, 2 in recovery	Gender: 10 male, 10 female Age: range 40–84; mean 57	8 deaths 2000–2009; six 2010-onwards; 6 in the 1990s and 1 in the 1970s.	17 deaths wholly or partially linked to long-term alcohol use; 1 involved both drugs and alcohol; 1 death involved methadone; 1 cancer and 1 accidental death in a fire
Spouses and partners (N = 13) Includes 10 partners/ spouses (1 LGBT); 3 ex-spouses/ partners	9 England, 4 Scotland	11 female; 2 male Age at interview: range 49–67; mean 56 Age at death 29–65; mean 45 2 in treatment, 4 in recovery	Gender: 10 male, 3 female Age: range 17–59; mean 45	6 deaths 2000–2009; 4 in the 1990s, 2 2010-onwards, and 1 in the 1980s	Four overdoses; 3 suicides; 3 linked to alcohol; 1 cancer; 1 hepatitis C; 1 described as 'sudden adult death syndrome'

Siblings (N = 12) Includes 7 sisters (2 step-sisters) and 5 brothers (1 half-brother) 1 brother talked about the death of 2 brothers	9 England, 3 Scotland	7 female; 5 male Age at interview: range 23–63; mean 43 Age at death: 20–58; mean 37 3 in treatment	Gender: 10 male, 3 female Age: range 23–61; mean 36	9 deaths 2000–2009 and 4 2010-onwards	4 related to alcohol; 3 overdoses; 1 suicide; 1 murder; 1 road accident; 1 cancer; 1 misadventure; 1 a result of drug use
Friends (N = 6)	5 England, 1 Scotland	5 male, 1 female Age at interview: range 30–49; mean 42 Age at death: range 23–40; mean 31 2 in treatment, 2 in recovery	Gender: all male Age: range 16–40; mean 29	4 deaths 2000–2009; 1 in the 1990s; 1 30 years ago	Four overdoses (1 suicide); 1 tuberculosis; 1 a result of alcohol
Extended family members (N = 3)	2 England, 1 Scotland	All female (nieces) Age at interview: range 24–40; mean 34 Age at death: range 19–38; mean 31 1 in treatment	Gender: 2 female, 1 male Age: range 45–80; mean 58	2o deaths 2010-onwards and 1 2000–2009	All 3 deaths associated with alcohol (also drugs in one case)
Treatment or recovery (N = 21) 7 spouses/partners (including ex-), 6 children, 4 friends, 3 brothers, 1 niece and 1 father 2 people talked about 2 deaths	15 England, 6 Scotland	12 male, 9 female Age at interview: 30–62; mean 44 Age at death: 15–56; mean 34	Gender: 15 male, 6 female Age: range 17–73; mean 41	10 deaths 2000–2009; 6 in the 1990s and 6 2010-onwards; 1 in the 1970s	7 overdoses; 3 suicides; 10 linked to alcohol; 1 classed as 'sudden adult death syndrome'; 1 murder; 1 cancer

(Continued)

Table 5.1 (Continued)

Group	Location of interviewees	Interviewee characteristics	Characteristics of the deceased	When the death occurred	Cause of death
Country (N = 106)	71 England (30 parents, 19 children, 9 partners, 9 siblings, 5 friends, 2 nieces) 35 Scotland (26 parents, 4 partners, 4 siblings, 2 children, 1 niece, 1 friend)	**England** – 49 female, 22 male Age at interview: 22–75; mean 51 15 in treatment/ recovery **Scotland** – 30 female, 5 male Age at interview: 23–75; mean 54 6 in treatment/recovery	**England** – 48 male, 18 female, mean age 41 (range 16–84) **Scotland** – 31 male, 5 female, mean age 33 (range 16–80)	**England** – Nearly half of deaths 2000–2009; approx. one-third 2010-onwards; 13 in the 1990s and 1 in the 1970s. **Scotland** – 18 2000–2009; 11 2010-onwards; 8 in the 1990s and 1 in the 1980s.	Exact CoD unclear in some cases or involved more than one substance. However, broadly, in **England** there was more variation, including more alcohol-related deaths and more suicides. In **Scotland** more than half of deaths involved a drugs overdose with many others involving complications associated with drug use.

¹In some cases an interviewee talked about more than one death, or more than one interviewee talked about the same person who had died. These data affect the numbers reported in this table. Also, data regarding substance type have not been included in this table because in some cases the information was unclear or involved more than one substance meaning that the substance[s] involved could not be easily quantified.

Robson and Walter, 2012; Wertheimer, 2001). Wertheimer's study of 50 family members, including 21 parents, bereaved by suicide suggests some particular features that are consistent with studies of parental bereavement more generally (see, e.g., Nadeau, 1998 and Riches and Dawson, 1996, 2000). These include the death breaking the family line and parental identity and being against the natural order of life; the death being accompanied by the loss of hopes and dreams which the parent had invested in the child's future; the death affecting relationships with any surviving children; and parents questioning what they did wrong and how they failed their child. Although there has been little research with parents bereaved by a child's substance-related death, Wertheimer's (2001) study of parents bereaved by suicide includes the further feature of direct or perceived stigma as a result of that 'failure'. Also, Feigelman et al.'s survey (2011) comparing parents bereaved through drugs, suicide, natural causes and accidents found greater stigma and less compassion for parents bereaved by drug-related deaths.

Parents were the largest group of interviewees in our sample ($N = 56$), particularly in Scotland, most often comprising mothers grieving a son's death. It is possible that this was due to recruitment strategies across the research sites, to be discussed later. Approximately half of the fathers who participated were interviewed together with their wives. Along with bereaved friends this was the group of interviewees where the average age of the deceased was lowest, at 29 years.

The aspects of interviewee experiences which appeared to be stronger in this group dovetail with those identified earlier in Riches and Dawson's and Wertheimer's study. So parents in our sample had a strong sense that their child's death went against the natural order of things and that they had somehow failed their child and hence failed as a parent, or that others would see them as 'bad parents'.

> If I had to lose him I wish it hadn't been through drug addiction because it makes you sometimes feel that you failed your child.
>
> (MotherE)[1]

> You do have a lot of guilt . . . you do feel I was a bad parent . . . people must think you are a bad parent . . . they must think well what made him turn to that.
>
> (MotherS)

Similarly, parents told us how they felt the death had affected other children and grandchildren, and the parents' relationships with other surviving children.

> Sometimes I feel guilty that I am not a good mum to [youngest son] because I am so wound up in my grief for [son who died . . . I am not the same mum to him that I should have been to him.
>
> (MotherS)

> *[My daughter] said I lost my parents when he died. I said have we changed that much? She said you have mum.*
>
> (ParentsE)

There also appeared to be some ways in which the bereavement experiences of parents in our sample were different to, or stronger than, other sub-groups in our sample, particularly when their child had died because of illegal drugs (see Templeton et al., 2016b). These parents had often been unaware of the severity of the problem, with some believing that the use was 'normal' and experimental or that the problem was absent or not serious because their child did not conform to drug taking stereotypes. Hence they realised too late how serious the problem was. In addition, in our sample, only parents experienced the distress of being the first person to find the body (this was the case for six parents), often were very angry (particularly when they believed others to be involved in the death in some way) and needed answers.

> *I think you think people who are really down in their luck, maybe begging in the streets, no money for clothes or barely enough for food, those weren't his circumstances at all.*
>
> (ParentsE)

> *Then you looked in on [our son] and couldn't wake him and you screamed down the stairs that . . . you couldn't wake [him] so I ran up, you phoned 999 while I tried to resuscitate him.*
>
> (ParentsE)

> *I needed someone to hate and he got my focus because of what he done . . . [as time went on] I started to let the anger go.*
>
> (FatherS)

A very few interviewees, all mothers, admitted that things had been so bad that they had contemplated leaving their son to die, with one believing that it would be better if they or their child were dead, and another querying having her son at all.

> *The best thing to do was finish him off . . . he's going to die anyway. Nobody's listening, nobody's helping us. Our whole family was in crisis.*
>
> (MotherE)

> *If you knew what you were going to face . . . honestly you would shoot them or you would shoot yourself.*
>
> (MotherS)

*I had a lovely boy and a great life with him but at that time the pain was horren-
dous. I would have rather not have had him than gone through that pain.*

(MotherS)

In summary, in addition to dealing with the grief associated with the loss of
a child, parents had to deal with deaths that went against the natural order of
life and led many to question their role as a parent. For a small number their
experiences of living with their child's alcohol or drug use led them to express
very strong emotional views about their child's life.

Children

*The debilitating factor for me was the absence of a father, or the absence of a supportive
family structure, in a sense that was the death, that was the loss, and so the actual physical
death of a human being was not as significant as that absence.*

(SonE)

There is a wealth of evidence for how children of all ages can be affected
by parental substance use (e.g., Adamson and Templeton, 2012; Barnard and
McKeganey, 2004; Velleman and Orford, 1999) and by the death of a parent
(e.g., Legg, 2015; Worden and Silverman, 1996). However, there has been a
dearth of research which has specifically considered children bereaved by a
parent's death as a result of alcohol or drugs. Two recent small UK studies have
offered some suggestions for what the particular impact on children might
be (Grace, 2012; Legg, 2015), with Legg suggesting that bereaved children of
alcoholics experience more stress and strain, and less support, than children
bereaved by other means. Wertheimer (2001) considered what might be particular in the experiences of the children she interviewed who were bereaved
by suicide compared with her other groups of interviewees, listing increased
rates of anxiety, anger, shame and behavioural difficulties; being shielded from
the full truth of the death and excluded from key events and decisions following the death; a sense of responsibility for the death; taking on a family caring/
parenting role; feelings of hurt, rejection and abandonment; and difficulties in
building trusting relationships with others.

Grown-up children were the second largest group in our sample ($N = 21$),
and all except one talked about the death of a parent to alcohol use. Most of this
group were adults when their parent died but some were teenagers (aged 19
and under) and one was a child of 9. Interestingly, those bereaved as a child or
young person tended to highlight the death of a father, whereas those bereaved
as an adult tended to discuss the death of a mother. There are examples in
our sample of children which mirror all the findings from Wertheimer's study
(2001) about how children can be particularly affected by a parent's suicide, but
also some additional features discussed next.

The main way in which this group appeared to differ from others in our sample was in the sense of disappointment at an unfulfilled parent/child relationship. In this regard they very much mirror other literature in this area. The parent's substance use had undermined or even destroyed their ability to behave as a parent might be expected to behave, cheating the child of a conventional relationship with the parent; for some interviewees this loss was greater and had a greater impact than the death itself.

> She was in the high dependency unit and I thought she was going to die and she didn't but the doctors basically said she will never be the same again ... in all honesty, I felt like that was the point at which I really lost her. I went through a grieving process then, which I think is probably more akin to what people would usually experience when they lose somebody, compared with what I went through when she actually died, which was completely different.
>
> (DaughterE)

Although some experienced loving relationships with their parents, this was often undermined by their parent's inability to care for them and provide a stable, carefree environment for their formative years. For some, this extended to mental, physical and emotional abuse at the hands of their parents, bringing extreme distress. The typically negative nature of interviewees' relationships with their parents when they were alive was for some compounded by their death, often accompanied by a loss of hope for the parent and the relationship they had lacked or lost, or sadness that parents would not be able to form relationships with their own (i.e., the interviewee's) children.

> I wasn't just grieving for the fact that my whole life had changed, it was also grieving for the father that I'd never had ... at least when he was alive, even when he was being absolutely awful there's always a tiny bit of me that thought well maybe one day I'll get the dad that everyone else gets ... when he died it was like well that's it, there is no more, there's never going to be another chance.
>
> (DaughterE)

This group also differed in the nature and intensity of their sense of responsibility and guilt. Some interviewees experienced guilt over their parent's death and its causes, such as a woman whose elderly mother died following a fall in her home and another whose mother had a problem with opiates.

> She had broken her femur but she had been on the kitchen floor for a week and she went into (hospital) on the Saturday and she died on the Sunday ... They asked me about her drinking and I suppose why I'm doing this [interview] is because it's the guilt in me. I feel that she died whereas I could have probably helped her a lot more.
>
> (DaughterE)

It felt to me like it was ultimate failure on my part . . . I feel a great sense of disappointment that having invested 15 years of my life into trying to help her to become strong . . . that it's ultimately failed.

(DaughterE)

On the other hand, some of this group had a level of detachment from any responsibility towards their parent, possibly reflecting expectations of conventional child/parent relationships where responsibility is expected to flow from parent to child rather than vice versa or because they had suffered enough and realised that ultimately there was nothing more they could do.

It's always that thing of 'I could have done something, I could have done something more' is what I hear from a lot of people. I didn't feel that because I knew there was nothing else more we could do.

(DaughterE)

In summary, what seemed to distinguish the bereavement experiences of this group was the loss of hope for regaining (or, for some, achieving for the first time) a conventionally loving and supportive parent–child relationship, which for many had been absent while their parent was alive. However, this was one of two groups in our sample where most interviewees were bereaved by an alcohol-related death (possibly a feature of our recruitment, where most interviewees in this group came from a charity which supports children affected by a parent's alcoholism) and the experiences discussed earlier may well reflect this.

Spouses/Partners

It was like his drinking was in control of both of us.

(PartnerS)

Historically, there has been relatively more research with partners/spouses who have been bereaved by the death of a spouse (e.g., Parkes and Prigerson, 2010; Stroebe et al., 2007) or who live with a substance-using spouse (e.g., Velleman et al., 1993; Velleman et al., 2007). With regard to bereavement, Wertheimer (2001) noted that the particular features of spousal bereavement by suicide were the practical impacts, including raising children, finances, isolation and loneliness; feeling rejected by their partner because they died; feeling responsible for the death; balancing their own needs with those of children which could put their own grief on hold; and reflecting on their choice of later partners.

In our sample, most of the 13 spouses/partners are female and comprise both current and ex-partners/spouses (including a same sex couple). Nearly half were in treatment or recovery at the time of the interview, and five had a history of using substances with the partner who died.

As with the other groups in our sample, there is considerable overlap between our findings and Wertheimer's on bereavement after suicide (2001), and little over and above these which stood out as more prominent for spouses/partners bereaved by substance use. One potential difference was that some felt excluded, for example, from decisions about the deceased's funeral or disposal of remains. This was more common for those who were not married to, or had separated/divorced from, the person who died. For example, one ex-wife did not know what had happened to her ex-husband's ashes, and another partner said that although her partner's family did include her in funeral arrangements they refused for a while to give her the ashes.

> They've had nothing to do with me almost since the day he died.
>
> (PartnerE)

> I would have liked different music [at the funeral], I was kind of overruled by his family in what is appropriate.
>
> (PartnerE)

> And because we'd already split up . . . I didn't really have my place in the funeral . . . so we were like one row back and it was very obviously one row back, there's a kind of protocol isn't there at funerals?
>
> (Ex-wifeE)

Furthermore, some partners decided not to attend the funeral; one interviewee sensed that others blamed her in some way for the death, and another wanted to avoid conflict with other mourners (see Chapter 4). For some, however, a degree of removal was a relief and reduced stress; one an ex-partner said that the police were considerate in asking whether they should discuss things with her or the deceased's brother and that the family did give her 'status' as the wife at the funeral.

> They gave me my place as his wife but didn't push anything on me.
>
> (Ex-partnerS)

In summary, this was the group for whom there were few clear ways in which their experiences of bereavement differed from other groups in the overall sample. One feature which seemed to be more prominent was the extent to which a separated partner or divorced spouse felt included or excluded in decisions and processes after the death. It could be inferred from our interviewees that non-blood relatives can be excluded or sidelined and that experiences can also be influenced by disagreements between blood and non-blood relatives. Overall, it is possible that differences for this group are more subtle or that there was more diversity within this group, making it harder to identify any unique characteristics of their bereavement.

Siblings

[My son said to me] Ma I've lost three brothers, I didn't think of it that way, I thought it was just me that had lost these boys.

(MotherS)

There has been a lack of research with bereaved siblings, particularly adult siblings affected by a brother or sister's substance misuse (Mash et al., 2013; Segal et al., 1995; Wertheimer, 2001; Wright, 2016; Zampitella, 2011). However, the research which is available suggests that the bereavement experiences of adult siblings (as in our sample) are very different from siblings bereaved as children (Wright, 2016) and are not recognised by society (Marshall and Davies, 2011; Wright, 2016; Zampitella, 2011), with Wertheimer describing bereaved siblings as 'forgotten victims' (2001) – aptly demonstrated by the quote earlier from a mother. Overall, Wright's (2016) review of adult sibling bereavement concludes that this is a particularly difficult loss, because of the nature of the sibling bond – often the longest relationship that an individual will have with another relative/person (see also Packman et al., 2006; Zampitella, 2011) but one which is largely misunderstood, minimised and disenfranchised (Marshall and Davies, 2011; Segal et al., 1995; Wright, 2016; Zampitella, 2011). An American study with 73 bereaved young adults, 7 following the death of a sibling (the others were friends), found that sibling loss was 'particularly distressing'; the stronger the sibling bond, the greater the psychological and physical symptoms (Mash et al., 2013). A number of studies have considered what might be particular to sibling bereavement, including strained relationships with remaining family; suppressing grief that is believed to be lesser than that of parents and hence feeling excluded by parental grief; unresolved, delayed or minimised grief; major health problems; strong emotional reactions including a desire for revenge; being excluded from key processes and events related to the death; taking on the role of a parent; survivor guilt and wondering if they did anything to influence the death; finding it difficult to move away from the family and be independent; and a lack of specific support (Marshall and Davies, 2011; Wertheimer, 2001; Wright, 2016; Zampitella, 2011).

There were almost equal numbers of male and female siblings in our sample, including step and half-siblings, although most were bereaved by the death of a brother. Their experiences confirm many of the findings summarised earlier, most notably strong emotions and grief, hiding or delaying grief to protect self and others (particularly parents). However, some of previous research findings are drawn from young rather than adult siblings. In a significant minority of our siblings, including brothers with a history of using substances (and engaging in other activities such as crime) together, there was evidence of a very close sibling bond.

Losing my sister was the worst thing that's ever, ever happened to me and I think barring the loss of my daughter or [my husband] I don't think I could have anything worse and to lose her the way we did was just the most awful thing that's ever happened in my life.

(SisterE)

We were always together, every day we would be together . . . we'd go out shoplifting together, even just hanging about together. We would be together all the time.

(BrotherE)

No-one knew her like me, certainly no-one knew me like her. And so much of my outlook and what I am into and my thing has been defined by her.

(BrotherE)

Where siblings have shared difficult experiences from growing up together, including parental substance misuse, the loss of a sibling can take on a particular significance. This confirms Wright's research which found that "when an adult sibling dies, the surviving sibling loses not only a confidant and companion but also connections to memories and events that are otherwise inaccessible". (2016, p. 35)

A bit of survivor's guilt maybe because we both had that awful childhood . . . I've lost the one person who I could have talked to about what it was like growing up with Mum and Dad. I haven't got her to go to, to say do you remember that Christmas or do you remember that family holiday, there's none of that . . . I don't have that person to share childhood memories, our pets or whatever. . . . there is also that loss of a shared history . . . I think that's probably the effect of losing a sibling over a parent or any other relative through alcoholism.

(SisterE)

Interestingly, stigma was not a dominant theme for this group, which warrants further exploration. In summary, the data confirm that the sibling bond, sometimes including shared experiences of substance use or difficult childhood experiences, is a unique one which can have wide ranging impacts on those left behind.

Friends

I just felt different because on paper I wasn't a family member.

(FriendS)

There has been a dearth of research into the impact of the death of a friend even though friendship bonds can lead to particularly difficult bereavements (Mash et al., 2013; Sklar and Hartley, 1990). It has also been suggested that social expectations and norms for grief are lacking for this group, making it hard for bereaved friends to access support. Sklar and Hartley (1990), who conducted the first known studies on 'survivor friends', involving a total of 45 adults, found that friends experienced "considerable unresolved feelings of despair, guilt . . . fear for one's own mortality, and a sense of emptiness that had not been resolved since the death" (1990, p. 108). They also found many parallels between the bereavement experiences and what is known about other groups of bereaved people. Sklar and Hartley (1990) also noted that such bereavements can be particularly difficult where the friend is seen as an extension to or substitute for the family. More recent studies have also concluded that people can be greatly affected by the death of a friend and to try and cope with their grief may increase their use of alcohol (Creighton et al., 2016). Another study of complicated grief in young adults, involving 66 friends and seven siblings, found that, although siblings' grief was greater, both groups had higher levels of depression and somatic symptoms than a comparison group of non-bereaved adults (Mash et al., 2013).

We interviewed six friends, five males in England and one female in Scotland. Two were in residential treatment at the time of interview and another two had been in recovery for a number of years (all in England). Five of the six experienced the death of a male friend between 1990 and 2007, whereas in the sixth case the death had taken place over 30 years previously. The majority of the deaths involved drug overdoses.

All six described very strong friendships, which were very important to them and which for some were stronger than relationships with their own family. One friend said that his friend's death had a far greater impact on him than the death of his father (who also died through substance use)

> We were like brothers, really. I'm closer to [him] than I am to my own brother . . . when he died that made me realise how significant the relationship was and I've never had another friendship like that.
>
> (FriendE)

> [It was] an important friendship . . . for three or four years he was like family, there was no-one else really.
>
> (FriendE)

Four male friends described a history of drinking and using drugs (and in two cases also dealing) with their friend, over some years and in some cases since they were teenagers.

> It was experimenting, exploring and we kind of just related to each other . . . I guess on a friendship level there was a connection . . . he was my main relationship at the time.
>
> (FriendE)

> I grew up with [my friend] for such a chunk of time, and we went through adolescence . . . we kind of found our identities together so when he died it felt [like] I was

robbed of part of my identity . . . it's really difficult to describe the loss because it was like part of me had died as well.

(FriendE)

Yet despite being particularly close to the person who died, they felt that, as a friend, their friend's family assumed that they were less affected by the death.

I had all these feelings [that] I wasn't actually part of the family.

(FriendE)

I was as close to family as you could probably get . . . but because I wasn't family no-one else recognised my loss . . . I don't think they stood back and thought, gosh, it must be really hard for me. And it's something that I was never ever able to say to them . . . I just felt different because on paper I wasn't a family member.

(FriendS)

Some went further to convey how they themselves had assumed that, as a friend rather than a family member, their own grief was less important.

I think because I wasn't family I thought I should do more for them because the family were all suffering so much . . . Probably I stood back a bit from the family and thought they should have their own grieving space . . . I think on reflection I didn't grieve for myself.

(FriendS)

Four interviewees felt guilty that they had not been able to do more to help their friend; for one this was very much tied up with his own drug problem. Another believed himself to be solely responsible for the death; he served a prison sentence for supply of drugs and manslaughter and more than five years later has not been able to come to terms with this, describing several overdose/ suicide attempts. During the interview, this man expressed his remorse.

If I could swap places in a second I would . . . I relive it every day . . . I think I will be forever trying to make up for it . . . I'm extremely sorry for what I've done . . . I think it's the first time I've said it aloud.

(FriendE)

Other differences for this group were that stigma was not a dominant issue and that interviewees were not involved with official processes such as post-mortems and inquests. They were therefore spared the stress that comes with stigma, post-mortems and inquests and that exacerbated the grief of many family members. This group also described mixed experiences of involvement in funeral arrangements, and of support, some suggesting that their needs are rarely considered.

There's no help for people with bereavement from friends that I know of.

(FriendE)

In summary, particular features of this group's experience include disenfranchised grief (Doka, 2002): feeling that their grief went unrecognised because they were not family, or not giving themselves permission to grieve because they were 'only friends'. Further, friends were more likely to be removed from some of the more stressful aspects of these deaths such as stigma and involvement with official processes. Overall, the strength and longevity of many of the friendships makes this a group whose grief is no less than that of other groups, but who feel that their grief is secondary. One mother recognised this, describing the friends who attended her son's funeral as "unseen mourners". Bereaved friends would therefore seem to be a group for whom specific support is not available and for whom accessing support is particularly difficult.

Extended family members

The need for family support needs to be very broad, it can't just be immediate family only.

(NieceE)

There is no known research which has considered the experiences of extended family members who are affected or bereaved by another's substance misuse. The extended family members in our sample comprised three nieces, two of whom were affected by an aunt's and the third by an uncle's death, all associated with alcohol. Given the lack of research in this area, do these three offer any new and valuable insights into bereavement by substance use? Being the smallest group in our sample it is difficult to identify what might be specific to nieces and what might be more widely representative of extended family members or, indeed, of any family member. Nevertheless, two things were of particular interest. First, all three nieces described ways in which the death had had a positive impact on them – one thought that it made a difference to her job, bringing her into contact with people (and their families) affected by substance use problems. For another it prompted her to focus on her career and her own mental health problems. The third niece expressed a desire to make renewed efforts to tackle her own alcohol problem.

Since my uncle died, I've turned my life around. I had nothing . . . my mental health was really quite severe. I was getting into lots of debt and, when he passed away, I just realised that you need to be living.

(NieceE)

It was a kick up the backside . . . when I relapsed [I] wanted to get help straightaway myself because I thought I am not going down that road and seeing her in hospital and knowing like everything gave up and mine was on its way again.

(NieceE)

Second, two nieces identified a lack of support for extended family.

> *I would have benefitted from immediate support . . . support also needs to be avail-*
> *able and to take into account the wider family . . . we didn't have any support . . .*
> *which I was so angry with . . . just because I'm not his partner or his child doesn't*
> *mean I'm not as close or as affected.*
>
> (NieceE)

> *The need for family support needs to be very broad, it can't just be immediate family*
> *only . . . actually all the extended family can be affected in a range of ways and that*
> *help needs to reach widely as well.*
>
> (NieceE)

In summary, the size of this group and the lack of other research with extended family members (regarding substance use and/or bereavement) make it hard to identify anything specific. Nevertheless, our data suggest that although stigma may not be dominant, extended family members can be greatly affected by the substance-related death of a relative; this may be neither recognised by others nor considered when it comes to support.

Those in treatment or recovery from substance use

> *Well, I haven't dealt with his death, hence me being an alcoholic.*
>
> (DaughterE)

Several risk factors, either on their own or in combination, can be associated with substance use. Bereavement, including, for example, the death of a parent in childhood or of a relative or drug-using peer later in life, is one such risk factor. A number of studies have suggested an association between the death of significant others, unresolved grief and either commencing or increased/ongoing problems (including risk of relapse) with substance use (Creighton et al., 2016; Denny and Lee, 1984; Masferrer et al., 2015; Stroebe and Stroebe, 1987; van Sydow et al., 2002). A Canadian study with 57 men found 54% used alcohol (and in some cases also other drugs) following the accidental death (in 11 cases as a result of alcohol or a drug overdose) of a male friend, most commonly to dull the pain or purge sadness (Creighton et al., 2016). However, substance users may not receive sufficient support for the bereavement. Denny and Lee reflected that many of their clients had not talked about their loss prior to substance use treatment but that it was important that they were "given permission to grieve and express sorrow without shame and guilt" (1984, p. 250). Professional bereavement help is often required alongside substance use treatment (e.g., Denny and Lee, 1984; Masferrer et al., 2015) because substance users may be isolated from or in conflict with their own family networks, or unwilling to burden their relatives more by talking about the deceased and their grief.

In our sample 12 were in treatment when they were interviewed and 9 described themselves as in recovery, in some cases for many years (Table 5.1) Most of these 21 were male and all talked about their own substance use history, which had usually lasted for many years and often co-existed with other problems. There was wide variation, however, in the relationship between their own substance use and that of the person who died. Roughly half described using alcohol/drugs problematically with the person who died and how this was a significant aspect of the relationship. The others used substances independent of the person who died. Just over half of this group were actively using alcohol or drugs when their relative or friend died, although there was variation in whether this was with or independent of the person who died.

Some expressed regret, sadness or guilt that their own problems with substances prevented them from offering support to others, including to the person who subsequently died or to other close relatives after the death.

> *I said I couldn't talk at that time . . . I was so wrapped up in my own addiction and withdrawals and kids and everything and I just said look give me a call in a couple of days and I will be able to talk . . . That's the sadness of addiction.*
>
> (FriendE)

> *I blame myself a lot, I blame myself that I wasn't there for him enough. He tried to get that relationship between us going and I didn't try hard enough . . . when I managed to stop drinking I felt responsible that I didn't help him.*
>
> (Adopted SonE)

Several described how, after the death, they continued (in some cases escalated) their substance use in order to avoid their own grief and bereavement.

> *I was stupid after he died with blocking everything out through all the drug abuse . . . I wasn't interested [in counselling], I just wanted to get out of my face.*
>
> (PartnerS)

> *[I] just used drugs to block everything out . . . and never really talked about it . . . It's been hard. I've learnt to live with it. By using heroin I've blocked it out.*
>
> (BrotherE)

> *I carried on for 18 months . . . running, running, running, trying to avoid the pain . . . it was probably the worst two years of my life . . . and of course when I put the drink down . . . I still had to start grieving again.*
>
> (SonE)

Others, however, reported how the death acted as a trigger to consider their own mortality and tackle their own substance use. This resulted in some being grateful for their own recovery and acknowledging the role of their friend's or relative's death in this; some were prompted to work, or train for paid or voluntary work, in drug/alcohol treatment services.

> *[I was] watching [my son] carrying his dad's coffin and I was sitting just looking at him and something came in and went you need to do something or you are going to be next and he will be left by himself.*
>
> (Ex-partnerS)

> *I'm in recovery and I feel really fortunate to be in recovery . . . not everyone gets here . . . it helps me hold that gratitude and it enforces that I don't want to use drugs. . .[my friend's] death was a waste and I don't want to be a waste.*
>
> (FriendE)

Inevitably, therefore, a clear focus for this group was their treatment and recovery, and how, for some, their involvement with treatment services enabled them to deal with their grief.

> *It's affected me a lot more than I am allowing myself to accept . . . I think talking about it now is making me aware that I have not dealt with it.*
>
> (FriendE)

> *I think I am really lucky that I've got close links here [treatment service] because I think really if I hadn't, if I had just been kind of flapping free, I think I quite likely would have gone back to using heroin by now.*
>
> (PartnerE)

> *It's only now, since I've been coming [here] that I've been offered grief counselling, and it's really helped in the last seven weeks I've [come] to terms with it more than I have in 18 years . . . If I had had the opportunity that I had here . . . if I'd had that years ago I would be such a different person, I would not have completely screwed up my life.*
>
> (DaughterE)

On the other hand others felt that not enough was done to include bereavement support within substance treatment and that it was something which needed to be handled carefully and sensitively, including the impact on other service users.

I thought in treatment they didn't look at it enough, I think they were trying to be too quick with me . . . because that's when all the stuff starts coming back, you know you feel all that stuff and take the drugs away from you, you start feeling all the stuff and they should be able to access you into something you know . . . when you get clean it just all hits you at once . . . you feel all the shame, all the guilt.

(Ex-partnerS)

In summary, interviewees conveyed a close association between their own sub-stance use and how they dealt, or failed to deal, with grief for their relative or friend. The death could trigger an escalation in or relapse to substance use, or be a catalyst for increased awareness of mortality and of subsequently entering treatment and maintaining recovery. It is therefore important that treatment and ongoing recovery includes bereavement support.

Substance type

Rates of both usage and mortality linked to alcohol or drugs are high across the UK, although there are some variations (see Introduction). For example, opiate use is much higher in Scotland; overall alcohol- and drug-related deaths are higher in Scotland than in England and Wales; both alcohol- and drug-related deaths in England are most common in the North compared with other regions, (ISD, 2016; ONS, 2014, 2016; United Nations, 2015). Despite such variation in rate of usage and of dying, families are often similarly affected by another's substance use regardless of the substance(s) involved (Arcidiacono et al., 2010; Orford et al., 2005; Velleman et al., 2013). Exceptions include one study of an intervention for family members which found that symptoms of ill health were higher for relatives of drug users than for alcohol users (Velleman et al., 2011). Perhaps the main differences families experience stem from the increased stigma of the illegality and uncertainty of drug use, compared with alcohol's legal status and social acceptability, though alcohol users are more likely to be violent and abusive.

Approximately one-third of deaths in our sample involved an overdose of drugs (usually heroin) and roughly a further quarter could be directly attrib-uted to alcohol. This prominence of deaths directly from heroin or alcohol reflects both usage and mortality data for the UK. In the other cases there was more than one substance involved or an accompanying factor such as sui-cide, illness or accident. This variation allows for some exploration of whether the substance(s) involved in the death in any way influenced our interviewees' experiences of bereavement.

In another publication we have considered 32 interviews associated with a fatal overdose (usually involving heroin) (Templeton et al., 2016b). This identi-fied several ways in which this group's experience differed from the rest of the sample – namely, occurrence of non-fatal overdoses before the death, which in some cases involved intervention from our interviewees, the sudden and often traumatic and brutal nature of the death, finding the deceased, knowledge or belief that someone else was involved in the death, having to negotiate complex procedures and official channels following the death (including how the body was handled) and the stigma linked to this particular death. In another publication analysing 14 interviews which involved an alcohol-related death (Valentine et al., 2016) we identified how with alcohol it seems that, before death, interviewees experienced before death more alcohol-related accidents and illnesses and longer drinking careers which had often lasted for many years. Thus, the deaths were more likely to be associated with alcohol-related illnesses and less likely to require a post-mortem or inquest.

Clearly, differences also exist as a result of the cultural and social acceptability of alcohol compared with drugs and their closer link to illegal behaviour. As discussed in Chapter 3, some thought that deaths involving illegal drugs (and more specifically injecting heroin) attracted greater stigma and more extreme stereotypes than deaths involving alcohol, and that this affected how they reacted and how they were treated after the death.

> I think taking heroin, smoking it somehow seems less desperate than injecting it . . . I guess there's almost a snobbery about it . . . somehow it seemed more acceptable.
>
> (MotherE)

> It's only because we were determined to see them both the same, drink and drugs, because they are the same, the effects are the same. The despair is the same . . . There is more dignity if you are an alcoholic by a long chalk. People treat you like you a human being and the hospitals treat you as a human being, but they don't when you are a drug addict, they treat you like you are the scum of the earth.
>
> (MotherS)

> I don't think there's as much stigma if it's alcohol . . . There's a kind of hierarchy. Heroin is always looked on as a dirty drug. Cocaine, while it can be just as damaging, it's seen as quite a clean drug . . . alcohol is legal and it's socially acceptable.
>
> (CoupleE)

Another apparent difference was that alcohol-related deaths tended to follow long-standing drinking careers, often accompanied by ill health. Although the death was not always expected interviewees and others often had the opportunity to say goodbye to or be with the person when they died.

> My uncle had been drinking since his teens . . . but it seemed to escalate over about 20 years . . . He went into rehab three times. The final time, they said in three years, you know you're going to be dead if you don't stop drinking, but he just couldn't . . .

and then three years later his liver just started to fail; his stomach started to swell. He went into hospital . . . and then seven days' later he died.

(NieceE)

Drug-related deaths, by contrast, tended to follow shorter drug-taking careers and occur suddenly (although not necessarily unexpectedly) in often traumatic and brutal circumstances.

[A]s I run into the hall [youngest son] was running down the stairs and he says to me [my brother] is not moving and he's not breathing, he is dead. And I just knew, I says to him he's not, he's not, he's just sleeping, but I knew and I said as I came in every footstep going up the stair, it seemed like slow motion and when I walked in the room and he was lying on his bed with his clothes on and he was just lying on his arm and his glasses on for reading . . . And he was all marked, all bruised and all marked.

(MotherS)

In summary, our data suggest some differences in the experiences of those bereaved by alcohol or illegal drugs. The differences derive from the nature of alcohol- and drug-using careers and from stigma, both of which interviewees believed to be more extreme when drugs were involved and more subtle and embedded in the case of alcohol. However, the predominance of heroin in our drug deaths sample may have influenced these findings; it is unclear whether such experiences are shared by people bereaved by the use of other drugs, including, for example, cocaine, ecstasy, volatile substances or novel psychoactive substances.

Geography

We have already seen that there are country and regional differences in both the use of substances and related mortality. It is likely that such differences will affect the experiences of those who are affected and bereaved by substance use. Other factors also need to be considered, including for example culture, social deprivation and religion. Although we did not collect data on all of these factors, recruiting interviewees from both England and Scotland provide a starting point for considering any differences.

There were some differences in the profile of our interviewees from England and Scotland (Table 5.1). Notably, although most interviewees were female in both countries, in Scotland three-quarters were parents (usually mothers who talked about the death of a son), whereas in England less than half of interviewees were parents and there was greater diversity in how interviewees were related to the deceased. The mean age of the deceased was also younger in Scotland, which is unsurprising given that Scottish participants were largely parents bereaved by the death of a child. It is possible that recruitment strategies in the two countries accounted for some of these differences. For example, in England interviewees were recruited from a wider range of sources, whereas recruitment in Scotland was mainly via community support groups which were more likely to be attended by parents.

Accounts from Scotland and England indicate similar perceptions of stigma and experiences of being stigmatised; however, they also suggest a subtle difference in the frame of reference to stigma, with interviewees in Scotland more regularly referring to the popular image of the illicit drug–using 'junkie' and their fears that their relative was identified as such.

> That was really difficult that this taxi driver had just labelled him a junkie, like that was ok: anybody that uses drugs is a junkie and deserves to die then, that's hurtful and ignorant and I think there's too much of that opinion in society.
>
> (BrotherS)

> I just felt as if [the police] had went in there, it's a hostel for drug addicts and prostitutes so basically I feel they went in there that morning and went 'Junkie' and walked back out . . . they thought my daughter was just another junkie in a junkie hostel, dead.
>
> (MotherS)

Although the stigma of the word 'junkie' was also highlighted in England, discussions of stigma were much less dominated by this kind of language and involved (relatively) more neutral terms such as 'addict' or feelings of shame or embarrassment.

> There was a time when [son] died, the police attitude at the time, I did feel like, do they think I am a criminal, do they think I am worthless, you can't help but have that go through your mind.
>
> (MotherE)

> The coroner's court, there was a person that we had to get in touch with . . . he told me he was an ex police officer. He was extremely disrespectful to [my daughter], he said we won't need you to come and look at the body because we have her records . . . and he said she's been a bit of a naughty girl in the past hasn't she?
>
> (MotherE)

Hence, there were concerns among interviewees about how the deceased's substance using behaviour affected others' perceptions. Such concerns were particularly pronounced in Scottish accounts of exchanges with professionals both before and after the death, for example, being seen and treated as part of an underclass characterised by fecklessness, immorality and ignorance.

Because you don't have your high class education as I call it. I think they think you are nothing that's really the way we were treated . . . I said don't you [procurator fiscal] dare speak to me like that . . . I didn't leave school with a trail of exam results behind me but I didn't sit in the corner with a pointy hat with a big D [Dunce – a stupid person] on it either.

(MotherS)

I remember . . . going away down to the Royal Infirmary to see him. And walking in well dressed, my husband in a shirt and tie . . . it was just awful, hospitals particularly treat you very badly when you are a drug addict or a parent of a drug addict . . . They treat you like you are a second class citizen.

(MotherS)

Another English/Scottish difference was more frequent reference to bereavement support from small, community-based family support groups as a source of bereavement support in Scotland. Although this may be partly a feature of recruitment in Scotland, it is also the case that such support was much less frequently identified in England. A large minority of Scottish interviewees, all parents and mostly mothers, had obtained crucial support in their bereavement by attending local support groups where they were able to meet and share their grief with others in similar circumstances. Attendance at such groups, although not all specifically targeted at the bereaved, provided safe and comfortable environments that allowed members to articulate the worst of their grief as well as to build strong, long-term, friendships and experience welcome light relief.

There is a bond with the girls – and if you are feeling bad they are there to listen. If you want to cry you can cry. If you want to laugh you can laugh. Some days we are all down and we are all crying together.

(MotherS)

Bereaved English parents also attended support groups, but not to the same extent, and, unlike in Scotland, few of the groups provided dedicated support for parents bereaved through substance use. This meant that there were sometimes tensions between members bereaved in different ways.

You've got people who've lost children through cancer . . . or fire, horrid things. Then you've got the suicide people, you know there are trains and just unimaginable things . . . And then you get the, 'them and us', because you'll get the mothers who; "well my daughter didn't choose to die did she?" And you're thinking well no but then I think my child was probably obviously mentally deranged at the moment to go and do what he did. I mean who would go and stand in front of a bloody train for Christ's sakes.

(MotherE)

We expected to find further differences between the two countries, particularly interviewee experiences of the police and of official investigations because of the differences in how these are legally structured and undertaken. In England, the focus is on the coroner, and inquests, if held, are usually public; in Scotland inquiries are conducted by the procurator fiscal, usually in private – in theory more family friendly. However, we did not find the anticipated differences, with interviewees in both countries describing mixed experiences of such investigations and whether or not they provided the answers they sought about the death.

In summary, the main ways in which the experiences of interviewees in England and Scotland appeared to differ were the experiences of stigma and the focus in Scotland on family support groups. It is unclear whether the latter is a result of recruitment or some other factor, for example, family support being organised differently across the two countries or because of country or regional differences in alcohol and drug use and associated mortality. Further, most Scottish interviewees were mothers, possibly masking any further differences between the two countries.

Conclusions

This chapter started with what is known about the core experience of those affected by the substance use of someone close (Chapter 1) and what our study has highlighted about the core experience of those who are bereaved by substance use (e.g., Templeton et al., 2016a) and then investigated how these experiences might vary when a number of variations within our sample are considered. This leads us to make four broad conclusions. First, although elements of the core experience are present in all the groups in our sample, we have identified how that experience may differ between groups and how some aspects of bereavement may be stronger within some groups than others. One which is particularly interesting in this regard is stigma. Stigma was prominent across the entire sample (Templeton, Ford, et al., 2016; Templeton, Valentine, et al., 2016; Walter et al., 2015), but it seemed more dominant for some groups (parents, partners, children) than others (siblings, friends, extended family members, those in treatment and recovery). Stigma also seemed strongest for deaths involving heroin (Lloyd, 2010; Ormston et al., 2010; Singleton, 2010). Finally, different components of stigma may be stronger for some groups than others, for example, the different language used and attitudes expressed by interviewees in England and Scotland.

Second, for some groups there are both overlaps with other research which has considered variation in bereavement (e.g., Wertheimer, 2001), but also some emerging ideas as to what might be the specific impact of bereavement through substance use for some groups. This includes under-researched groups such as extended family members, friends and those in treatment or recovery. Examples include parents feeling that they failed their child in both life and death, children losing hope that an often difficult relationship with their parent can

ever be reversed, friends (and possibly also extended family members and part-
ners/spouses) feeling that their grief is secondary to that of others, non-blood
relatives such as partners, spouses and ex-spouses feeling excluded from some
aspects of the grieving processes, such as the funeral arrangements, and those
struggling with their own treatment or recovery from substance use recognising that their grief is tied up with all of this. All these ideas require further
investigation.

Third, our sample is not necessarily representative of the fairly large population of adults who experience this type of bereavement and so further research
is needed to test the ideas proposed here. Further, there are groups bereaved
by substance use who are not well represented in our sample, such as those
from black and minority ethnic groups, those who are lesbian, gay, bisexual or
transgender and those affected by drugs other than heroin. Nor have we been
able to consider in this chapter gender, age, social class, or deaths from suicide
or murder. These are all areas for further testing of which are the core elements
of being bereaved by substance use as well as where there might be between-
group differences.

Overall our study highlights the complexity of understanding the experiences of adults bereaved by substance use. Diversity can be seen across our sample and within and between its various subgroups. It is also possible that some
of the ideas put forward in this chapter are associated with particular biases in
our sample concerning some groups, such as mothers who were more likely to
have experienced the drug-related death of a son, friends who were male (with
one exception), or children and extended family members who (again with one
exception) discussed an alcohol-related death. Also, when exploring diversity
and difference in bereavement it is important to look beyond the straightforward
closeness of the relationship between bereaved and deceased (Armstrong and
Shakespeare-Finch, 2011). This was also recognised by Mash et al. who said that
"the level of closeness and conflict between bereaved and deceased. . . [before
death] . . . may be important to the course of the grief response" (2013, p. 1203).
This is most clearly seen with the friends and extended family members in our
sample where the relationship between interviewee and deceased was often
very strong, making the loss as great or greater than that of other interviewees
with a closer familial tie to the deceased.

Our examination of diversity has clear implications for supporting people
bereaved by substance use. Support should cater for diversity and be more
widely available for siblings, friends, extended family members and anyone else
currently lacking formal support. There is also scope for offering more bereavement support as part of, or alongside care packages for those in treatment or
recovery for their own alcohol or drug use. To facilitate such developments, the
practice guidelines that evolved from our study could be a useful starting point
(Chapter 7; Cartwright, 2015).

This chapter has highlighted diversity as well as commonalities in the experiences of adults bereaved through substance use. We have suggested that diversity

takes many forms and considered a number of aspects present in our sample. We have thus proposed how the experiences of groups within our sample appear to diverge both from that of other groups and from the core experience which can be seen across our whole sample. The ideas we have put forward should be seen as a starting point for understanding the range of experiences and needs of this largely neglected group of bereaved people, further exploration being needed to inform appropriate support delivery.

Note

1   'E' and 'S' refer to interviews in England and Scotland, respectively.

References

Adamson, J., and Templeton, L. (2012). *Silent Voices – Supporting Children and Young People Affected by Parental Alcohol Misuse: A Comprehensive Literature Review*. London: Office of the Children's Commissioner for England.

Ahuja, A., Orford, J., and Copello, A. (2003). Understanding how families cope with alcohol problems in the UK West Midlands Sikh community. *Contemporary Drug Problems*, 30, 839–873.

Arcidiacono, C., Velleman, R., Procentese, F., Berti, P., Albanesi, C., Sommantico, M., and Copello, A. (2010). Italian families living with relatives with alcohol or drugs problems. *Drugs: Education, Prevention and Policy*, 17(6), 659–680.

Armstrong, D., and Shakespeare-Finch, J. (2011). Relationships to the bereaved and perceptions of severity of trauma differentiate elements of posttraumatic growth. *Omega*, 63(2), 125–140.

Barnard, M. (2007). *Drug Addiction and Families*. London: Jessica Kingsley Publishers.

Barnard, M., and McKeganey, N. (2004). The impact of parental problem drug use on children: What is the problem and what can be done to help? *Addiction*, 99(5), 552–559.

Cartwright, P. (2015). *Bereaved Through Substance Use: Guidelines for Those Whose Work Brings Them Into Contact With Adults Bereaved After a Drug or Alcohol-Related Death*. Bath: University of Bath. Available at: http://rebrand.ly/bereavementguidelines

Creighton, G., Oliffe, J., Matthews, J., and Saewyc, E. (2016). "Dulling the Edges": Young men's use of alcohol to deal with grief following the death of a male friend. *Health & Education Behaviour*, 43(1), 54–60.

Denny, G., and Lee, L. (1984). Grief work with substance abusers. *Journal of Substance Abuse Treatment*, 1, 249–254.

Feigelman, W., Jordan, J., and Gorman, B. (2011). Parental grief after a child's drug death compared to other death causes: Investigating a greatly neglected bereavement population. *Omega*, 63(4), 291–316.

Grace, P. (2012). *On Track or Off the Rails? A Phenomenological Study of Children's Experiences of Dealing With Parental Bereavement Through Substance Misuse*. Unpublished PhD Thesis, University of Manchester, UK.

Holland, J., and Neimeyer, R. (2011). Separation and traumatic distress in prolonged grief: The role of cause of death and relationship to the deceased. *Journal of Psychopathology and Behavioural Assessment*, 33, 254–263.

Information Services Division. (2016). *Estimating the National and Local Prevalence of Problem Drug Use in Scotland 2012/13*. Edinburgh: Information Services Division.

Legg, T. (2015). *Understanding the Experiences of Bereaved Children of Alcoholics*. Unpublished Masters of Science Dissertation, University of Bristol.

Lloyd, C. (2010). *Sinning and Sinned Against: The Stigmatisation of Problem Drug Users*. London: UK Drugs Policy Commission.

Marshall, B., and Davies, B. (2011). Bereavement in children and adults following the death of a sibling. In Neimeyer, R. (Ed.). *Grief and Bereavement in Contemporary Society: Bridging Research and Practice*. New York: Routledge. pp. 107–116.

Masferrer, L., Garre-Olmo, J., and Caparros, B. (2016). Risk of suicide: Its occurrence and related variables among bereaved substance users. *Journal of Substance Use*, 21(2), 191–197.

Masferrer, L., Garre-Olmo, J., and Caparros, B. (2015). Is there any relationship between drug users' bereavement and substance consumption? *Heroin Addiction Related Clinical Problems*, 17(6), 23–30.

Mash, H., Fullerton, C., and Ursano, R. (2013). Complicated grief and bereavement in young adults following close friend and sibling loss. *Depression and Anxiety*, 30, 1202–1210.

Nadeau, J. W. (1998). *Families Making Sense of Death*. London: Sage.

Office for National Statistics. (2016). *Alcohol Related Deaths in the United Kingdom: Registered in 2014*. London: Office for National Statistics.

Office for National Statistics. (2014). *Deaths Related to Drug Poisoning in England and Wales: 2014 Registrations*. London: Office for National Statistics.

Orford, J., Natera, G., Copello, A., Atkinson, C., Tiburcio, M., Velleman, R., Crundall, I., Mora, J., Templeton, L., and Walley, G. (2005). *Coping With Alcohol and Drug Problems: The Experiences of Family Members in Three Contrasting Cultures*. London: Taylor and Francis.

Ormston, R., Bradshaw, P., and Anderson, S. (2010). *Scottish Social Attitudes Survey 2009: Public Attitudes to Drugs and Drug Use in Scotland*. Edinburgh: Scottish Government Social Research.

Packman, W., Horsley, H., Davies, B., and Kramer, R. (2006). Sibling bereavement and continuing bonds. *Death Studies*, 30, 817–841.

Parkes, C., and Prigerson, H. (2010). *Bereavement: Studies of Grief in Adult Life*. Fourth Edition. London, New York: Routledge.

Riches, G., and Dawson, P. (2000). *An Intimate Loneliness: Supporting Bereaved Parents and Siblings*. Buckingham: Open University Press.

Riches, G., and Dawson, P. (1996). Communities of feeling: The culture of bereaved parents. *Mortality*, 1(2), 143–161.

Robson, P., and Walter, T. (2012). Hierarchies of loss: A critique of disenfranchised grief. *Omega: Journal of Death & Dying*, 66(2), 97–119.

Scaife, V. (2008). Maternal and paternal drug misuse and outcomes for children: Identifying risk and protective factors. *Children & Society*, 22(1), 53–62.

Segal, N., Wilson, S., Bouchard, T., and Gitlin, D. (1995). Comparative grief experiences of bereaved twins and other bereaved relatives. *Personality and Individual Differences*, 18(4), 511–542.

Singleton, N. (2010). *Attitudes to Drug Dependence: Results From a Survey of People Living in Private Households in the UK*. London: UK Drugs Policy Commission.

Sklar, F., and Hartley, S. (1990). Close friends as survivors: Bereavement patterns in a "hidden" population. *Omega*, 21(2), 103–112.

Stroebe, M., Schut, H., and Stroebe, M. (2007). Health outcomes of bereavement. *The Lancet*, 370, 1960–1973.

Stroebe, W., and Stroebe, M. (1987). *Bereavement and Health: The Psychological and Physical Consequences of Partner Loss*. Cambridge: Cambridge University Press.

Sydow van, K., Lieb, R., Pfister, H., Hofler, M., and Wittchen, H. (2002). What predicts incident use of cannabis and progression to abuse and dependence? A 4-year prospective examination of risk factors in a community sample of adolescents and young adults. *Drug and Alcohol Dependence*, 68(1), 49–64.

Tedeschi, R., and Calhoun, L. (2004). Posttraumatic growth: Conceptual foundations and empirical evidence. *Psychological Inquiry*, 15(1), 1–18.

Templeton, L. (2012). Dilemmas of grandparents caring for grandchildren because of parental substance misuse. *Drugs: Education, Prevention and Policy*, 19(1), 11–18.

Templeton, L., Ford, A., McKell, J., Valentine, C., Walter, T., Velleman, R., Bauld, L., Hay, G., and Hollywood, J. (2016a). Bereavement through substance use: Findings from an interview study with adults in England and Scotland. *Addiction Research & Theory*. Available at: http://dx.doi.org/10.3109/16066359.2016.1153632

Templeton, L., Valentine, C., McKell, J., Ford, A., Velleman, R., Walter, T., Hay, G., Bauld, L., and Hollywood, J. (2016b). Bereavement following a fatal overdose: The experiences of adults in England and Scotland. *Drugs: Education, Prevention & Policy*. Available at: http://dx.doi.org/10.3109/09687637.2015.1127328

United National Office on Drugs and Crime. (2015). *World Drug Report 2015*. United Nations publication, Sales No. E.15.XI.6.

Valentine, C., Templeton, L., and Velleman, R. (2016). There are limits on what you can do biographical reconstruction by those bereaved by alcohol-related deaths. In Thurnell-Read, T. (Ed.). *Drinking Dilemmas: Space, Culture and Identity*. London: Routledge, pp. 187–204.

Velleman, R., Bennett, G., Miller, T., Orford, J., Rigby, K., and Tod, A. (1993). The families of problem drug users: A study of 50 close relatives. *Addiction*, 88, 1281–1289.

Velleman, R., Copello, A., and Maslin, J. (2007). *Living With Drink: Women Who Live With Problem Drinkers*. London: Pearson Education.

Velleman, R., and Orford, J. (1999). *Risk & Resilience: Adults Who Were the Children of Problem Drinkers*. London: Harwood.

Velleman, R., Orford, J., Templeton, L., Copello, A., Patel, A., Moore, L., Macleod, J., and Godfrey, C. (2011). 12-month follow-up after brief interventions in primary care for family members affected by the substance misuse problem of a close relative. *Addiction Research & Theory*, 19(4), 362–374.

Walter, T., Ford, A., Templeton, L., Valentine, C., and Velleman, R. (2015). Compassion or stigma? How adults bereaved by alcohol or drugs experience services. *Health and Social Care in the Community*. doi 10.1111/hsc.12273.

Wertheimer, A. (2001). *A Special Scar: The Experiences of People Bereaved by Suicide*. 2nd Edition. Hove: Routledge.

Worden, J., and Silverman, P. (1996). Parental death and the adjustment of school-age children. *Omega*, 33(2), 91–102.

Wright, P. (2016). Adult sibling bereavement: Influences, consequences, and interventions. *Illness, Crisis & Loss*, 24(1), 34–45.

Zampitella, C. (2011). Adult surviving siblings: The disenfranchised grievers. *Group*, 35(4), 333–347.

Part II

Services

Christine Valentine and Linda Bauld

Part I's account of bereaved people's experiences following an alcohol- or drug-related death could probably be replicated in many parts of the world, whereas Part II's focus on how services deal with substance use deaths and those left behind is specific to the UK. Nonetheless, our findings on how services could better support this group do have international relevance, not only for services dealing with substance use deaths and bereavement, but also those dealing with death and bereavement more generally and with health and social care. As a result of structural reforms to service delivery, inter-agency working is now the norm in several countries, and there is a considerable body of literature on both the potential and the pitfalls of this approach to delivering public services. This literature has found that barriers to effective service delivery include insufficient resources, understanding, communication, access and sharing between services, as well as competing priorities (Robinson and Cottrell, 2005; Hall, 2005; Jones, 2006; Kvarnström, 2008 and Milne et al., 2015). It has also found that the emphasis on structural reform of service delivery has been at the expense of responding compassionately to service users (Ballatt and Campling, 2011; Cole-King and Gilbert, 2011; Jones, 2013; Lown et al., 2012; Youngson, 2010).

The challenges of inter-agency working more generally were reflected in our focus group discussions about services dealing with substance use death and bereavement. In presenting the findings from these discussions we make reference to a 'services map' (see Figure 6.1). By powerfully conveying the scale, complexity and fragmentation of the 'system' that bereaved people must negotiate at a time when they are at their most vulnerable, this map represents one of our key findings. By considering the implications of existing provision for both bereaved people and practitioners, the following two chapters identify inadequacies and gaps as well as good practice. These in turn shaped the guidelines we developed to promote and support good practice, as follows:

- Chapter 6 examines existing service provision for those bereaved following a substance-related death. This examination is based on an analysis of data from six focus group discussions involving 40 practitioners from a range of services involved in dealing with substance use deaths, as well as relevant data from the interviews with bereaved people, which we reported in Part I.

- Chapter 7 presents five key messages for service improvement which we identified from both the interview and focus group findings. It discusses how these messages informed a working group of 12 members. Including both practitioners and bereaved people, the group was tasked with developing best practice guidelines that are accessible to a range of services and could be embedded in existing practice.

The final chapter draws together the threads of both Parts I and II, further underlining the value of research that directly engages key stakeholders in developing and implementing the study's findings:

- Chapter 8 summarises the studies' findings. It draws together the various threads to provide an overview and assessment of the study's achievements, including the relationship between the findings and the methods used to obtain them. The chapter also considers how these methods enabled the five key messages to be translated into guidelines that can be embedded in existing practice. It considers the implications of both the methods and the main findings of the study for this group of bereaved people and how engaging representatives of relevant services has enabled the development of guidelines that are both evidence and practice based.

References

Ballatt, J., and Campling, P. (2011). *Intelligent Kindness: Reforming the Culture of Healthcare.* London: Royal College of Psychiatrists.

Cole-King, A., and Gilbert, P. (2011). Compassionate care: The theory and the reality. *Journal of Holistic Health Care,* 8(3), 29–37.

Hall, P. (2005). Interprofessional teamwork: Professional cultures as barriers. *Journal of Interprofessional Care,* 19(Supplement 1), 188–196.

Jones, P. (2013). Equality of care: Substance users, the emergency care perspective. *Journal of Paramedic Practice,* 5, 548–549.

Jones, A. (2006). Multidisciplinary team working: Collaboration and conflict. *International Journal of Mental Health Nursing,* 15, 19–28.

Kvarnström, S. (2008). Difficulties in collaboration: A critical incident study of interprofessional healthcare teamwork. *Journal of Interprofessional Care,* 22(2), 191–203.

Lown, B. A., Rosen, J., and Marttila, J. (2012). An Agenda for improving compassionate care: A survey shows about half of patients say such care is missing. *Health Affairs,* 30, 1772–1778.

Milne, J., Greenfield, D., and Braithwaite, J. (2015). An ethnographic investigation of junior doctors' capacities to practice interprofessionally in three teaching hospitals. *Journal of Interprofessional Care,* 29(4), 347–353.

Robinson, M., and Cottrell, D. (2005). Health professionals in multi-disciplinary and multi-agency teams: Changing professional practice. *Journal of Interprofessional Care,* 19(6), 547–590.

Youngson, R. (2010). Taking off the armor. *Illness, Crisis & Loss,* 18, 79–82.

Dealing with substance-related deaths

*Jennifer McKell, Christine Valentine
and Tony Walter*

Substance-related bereavement has received little research attention, but even less has been paid to how practitioners and services respond. This reflects a wider dearth of research into the responses of practitioners to bereavement generally (NCPC, 2014). Where research does exist, the focus has been on the function of a country's 'death system' for the country as a whole, the implication being that in general 'the system' works (Kastenbaum, 2007). There is evidence from the UK, however, that bereaved people as a whole are poorly served by the death 'system' and often face gaps and inconsistencies in service delivery (NCPC, 2014). Such gaps and inconsistencies have also been identified in other areas of service delivery, particularly health and social care, although few studies have considered the impact on service users (Atkinson et al., 2007). Research has found significant service failure in responding to people bereaved by suicide (Peters et al., 2016; Chapple et al., 2013; Biddle, 2003), but no previous research has focused on services that deal with substance-related death and bereavement.

Our study of substance-related bereavement found that those affected face additional difficulties as they encounter 'the system'. They report confusion as to official procedures and mixed, sometimes insensitive, responses from a range of practitioners (Chapter 2; Valentine and Bauld, 2016; Valentine et al., 2016). As well as setting out bereaved people's experiences, this study has elicited practitioners' experiences of and views on substance-related bereavement and the response of services. Focus group discussions with practitioners working in England and Scotland helped us understand first the complexity of the system which both bereaved people and practitioners have to navigate, described in the first part of this chapter; and second the challenges faced by practitioners, often linked to specific working environments, presented in the second part.

Throughout, we place 'the system' in quote marks because its fragmentation is such that, in truth, it is not one system; it is perhaps better described as very complex terrain which practitioners have to navigate without a decent map and through which bereaved people must find their way usually without any map at all. Chapter 2 presented bereaved people's experiences of particular services; this chapter views the overall terrain as a (fragmented, confusing) whole.

A fragmented, confusing 'system'

Responsibilities for dealing with substance-related deaths and with the family and friends left behind are split across a range of services and organisations that can be broadly categorised into those that focus on the deceased and those that focus on the bereaved:

i) Services focusing on the deceased

Procedures required by law include establishing the cause of death. This may involve paramedics, the GP, the police and the coroner (in England) or procurator fiscal (in Scotland) and the pathologist. As with other cases of sudden, unexpected death (such as cot death or suicide), the home may be treated as a crime scene, the deceased's body and/or possessions taken into custody and the funeral delayed until after an inquest or ongoing police investigation. Such delays can create considerable uncertainty for the bereaved, who, as we shall see, may feel under suspicion as well as deprived of their family member's body or possessions. In addition, newspaper reporters may be tasked by their editor to produce a story about the deceased and the death. Practitioners whose work focuses on the deceased may easily sideline the bereaved, for a range of reasons discussed later in this chapter.

Other organisations – funeral directors, cemeteries, crematoria – are tasked with ensuring proper care and disposal of the body. Although working within a legislative framework of public health, these organisations, whether commercial or municipal, differ from all other practitioners (apart from clergy and funeral celebrants) in that they are responsible to and paid by the family.

ii) Services for those left behind

These include clergy or other funeral celebrants, bereavement counsellors and support groups, and family support groups and drug and alcohol services where the bereaved person is in treatment for their own substance use. Funeral directors are unusual in that they focus on caring both for the body and for their bereaved client.

A map

However, this neat categorisation of services and organisations is misleading. In practice, the post-death terrain is far more complicated, and is experienced as such by families. Figure 6.1 provides a map of the terrain. If at first glance it appears confusing, that is our intention because that is how families experience it.

Some of the organisations, services and people depicted on the map are encountered by families after any death. A doctor has to issue a death certificate; someone close to the deceased has to go to the local register office to register the death; friends, family and employer need to be informed; a funeral is arranged; and all manner of practical and financial issues have to be sorted out with utility companies, banks and so forth. Even after an 'ordinary' death, those taking responsibility for such matters can find them daunting. However, when

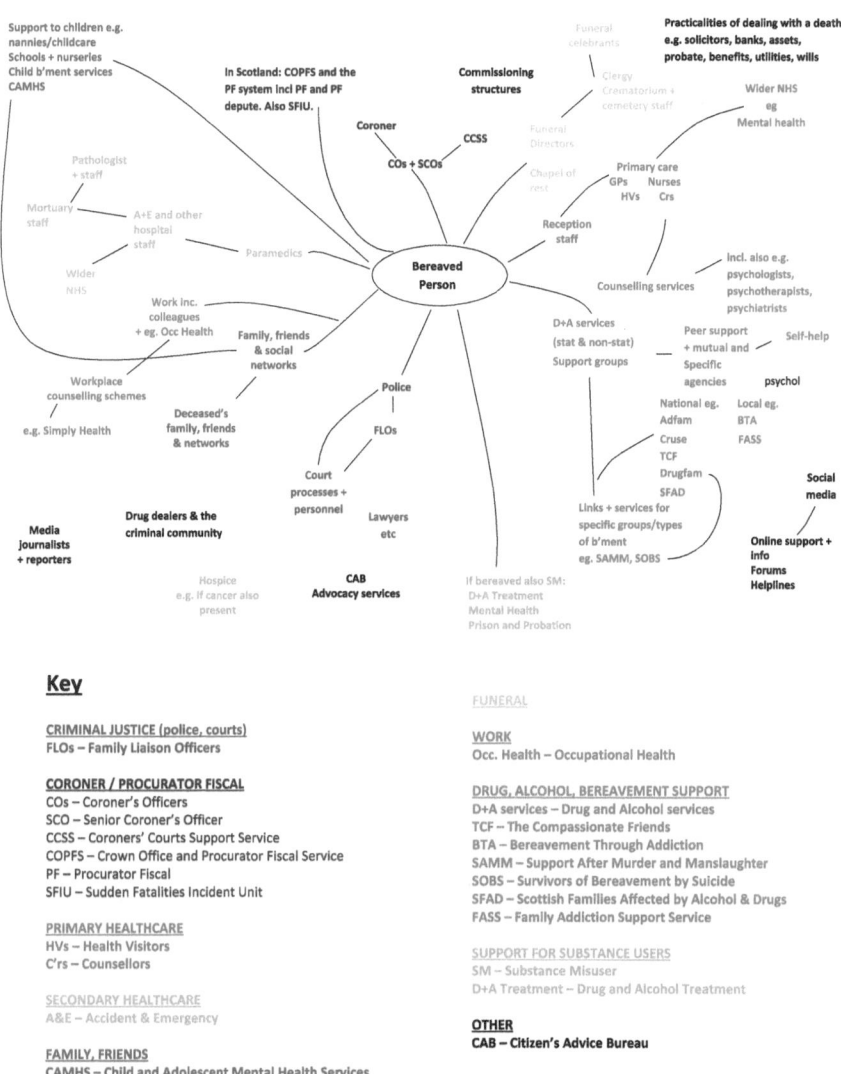

Figure 6.1 The terrain facing family members after a substance-related death

Note: Please see plate section for the colour version of this figure.

the death is alcohol related, and even more so when drug related or sudden, the terrain gets more complicated still.

Figure 6.1's use of colour coding offers a way in to finding some order within the complexity of who and what families encounter after a substance use death. First, members of the criminal justice system may be encountered (represented in purple). The family may be informed of the death by a police officer calling at the door, and if the death was sudden and at home then the family home may be treated as a potential crime scene. Second, if the death occurred in hospital, or the person died en route to hospital, then a range of health personnel will be encountered, from paramedics to accident and emergency nurses and doctors, to mortuary staff. Any post-mortem will be conducted by a pathologist who will produce a report (light blue). Third, these personnel are in addition to primary health care staff with whom the family are likely to already have been in touch (blue). Fourth, very probably the coroner service will be involved in England, or the procurator fiscal in Scotland (brown). Fifth, both the death itself and any subsequent investigation may well attract the attention of the local press (black). In the meantime, sixth, the family will be informing other family members and friends, and encountering – perhaps the first time – some of the deceased's friends (red) and if the deceased had been in work, seventh, informing the employer (grey). All the while, eighth, a funeral has to be arranged, entailing contact with funeral directors, a minister of religion or other celebrant (orange) and a range of other family members (red). Ninth, many of these communications will be employing digital and social media. In the long run, tenth, support may be sought from a range of bereavement services, whether generic or specific to substance-related bereavement (light green), and, eleventh, those bereaved who themselves use substances problematically may be in contact with treatment agencies, mental health services or the prison or probation service (green). Within each of these 11 areas, several organisations and several dozen individuals may be encountered. Complex terrain indeed.

Fragmentation

Not only is the terrain complicated but also, perhaps not surprisingly from what we know elsewhere of inter-agency working, it is fragmented. Some of the services are statutory, others not; some organisations are part of the criminal justice system, some part of the health service, some work for the local municipality, whereas others belong to profit seeking companies or voluntary groups.

> I think the challenge around services joining up mirror the issues around services joining up to support people before they die who have got multiple problems ... They don't have enough resources, or it's cultural ... The sense of the fact that this is a person with their own characteristics ... is lost a bit and actually joining it all up is super challenging.
>
> (Policy director for support service for families affected by substance use)

Later in the chapter we see how the focus groups described the challenges of linking all this up.

Not only are services and organisations complicated and fragmented, but also nobody knows the complete picture. A striking feature of our inter-professional focus groups was how, at the end of the formal discussion, participants gravitated to each other to exchange contact details. Not unusually the focus group provided the first ever opportunity for participants to meet face-to-face with representatives of other services, and sometimes the group made members aware of other services that they had hitherto been completely unaware of. Occasionally, during the focus group discussion, participants became aware of very different languages, for example those of the police and of psychotherapy.

As well as the complex mix of services and organisations, there is the complexity of the deceased's life. Death may reveal much about a life of which other family members had been unaware. One Scottish clergyman described his approach to drug related funerals:

> One of the real sadnesses about funeral services of an addicted person is that there will often be a whole range of the people that were closest with them which is their addicted pals and it's almost like a big dividing line right down that says don't even acknowledge these people, and one of the challenges I have is to overturn that and say tell me about the friends, let me speak to the friends.

A drug and alcohol recovery worker added:

> I had someone whose son died and she just wanted to come in and talk to get a sense of the sorts of things her son had been doing here before he died. We get approached quite a lot from families in this way.

Like these two practitioners, social media may also help fill in family members' gaps in knowledge the deceased's life; but social media may also do the opposite, announcing and pre-judging the death before the family have even been formally notified:

> It sometimes happens in the university world when there is a death of a student, people find out and parents sometimes or close relatives sometimes will find out because somebody has stuck it on Facebook or Twitter and that, and again that's when you get all the ill-informed speculation.
>
> (University chaplain)

Not only are services, families and social networks complex and fragmented, but the circumstances of each death and each family are different, so no one map would work for everyone.

> I agree with the complexity of the bereavement because of what it can dredge up for the person from the past. I think in dealing with this practically, you have to look at the circumstances around the person, because everyone's going to be so different.
>
> (Public health professional)

> Families have different attitudes to drugs – some may be involved in the person's treatment and others who want nothing to do with it, but will still be affected by the death.
>
> (Head of police drug strategy unit)

So not everyone wants or needs the same information. If the terrain is complicated, navigation tools would need to be tailor-made to the individual – a tall order in the shocking, stigmatising circumstances of some substance-related deaths.

Navigation

Chapter 2 showed how bereaved family members tried to navigate the system. By now it should be apparent how, even in the absence of stigma or insensitivity, this would be challenging. Although a number of register offices now have a 'Tell Us Once' service[1] which has significantly reduced the number of agencies a family has to contact in the event of an 'ordinary' death, there is no such system for the many additional agencies often involved in substance-related death. Moreover, there is no one official tasked with informing the family about the agencies they will have to contact and who will be contacting them. Though the police may provide a family liaison officer after a murder, terrorist attack or disaster, this is rare after substance-related deaths. As a senior coroner's officer said, referring only to the inquest process:

> I come from a very narrow focus in terms of supporting people when they attend the inquest process, but . . . when I talk to people the one thing that they say is that they have absolutely no idea about what to expect, what's going to happen, what the process will be and that's on top of trying to grieve.

What interviewees particularly appreciated was being kept updated and provided with explanations about what was happening and having their concerns listened to and taken seriously. However, for this to occur, individual practitioners need to be available and take time, with one person designated as a single point of contact. Such was the case for one mother who described the continuity of support she received from a police officer following her daughter's death.

> Well the DCI who'd been on the case right from the beginning, he was the one who'd gone in and . . . sorted out the room and everything, he was there and he was amazing, he was so good and he said "phone me any time" . . . If he wasn't on shift and I phoned up, they'd know straight away who I was, and they'd say, "Yes we will take a message" and he'd be back straight on the phone to me.

> (MotherE)

In contrast, as described in Chapters 2 and 3, encounters with practitioners who were not prepared to go beyond the immediate call of duty could undermine and alienate. Interviewees reported professionals and officials leaving them out of the picture at particularly crucial times, such as the point of death or its immediate aftermath.

Confusion

If lack of a map or a personal guide to the system would trouble even an unaffected navigator, it may deeply disturb family members already experiencing shock, anger or other powerful emotions often associated with grief. A sense that the world has gone mad can be a normal part of any bereavement, but certain forensic procedures after a sudden and unexplained substance related death can be additionally distressing to the bereaved. Focus groups highlighted bereaved people's experiences of forensic examinations led by police and forensic professionals following discovery of a body in the home. Concentrated activity by police and forensic staff intent on securing the environment and relevant items, including the body, in order to preserve and collect evidence can be experienced as an invasion of home and privacy:

> *At the time of death there is going to be lots of confusion ... you know, the Police coming in and so there is going to be a lot, people who are not going to be taking things in.*
>
> (Mortuary Services Manager)

Another layer of difficulty in such investigations is that the police are required to consider the possibility of criminal activity, making the home a crime scene and its residents and visitors potentially complicit:

> *We find that a lot of the families, some find themselves in a crime scene and when there is drugs in a house what often happens – the police they look upon the house, or the people in the house, as always complicit. But if your son has cocaine under his bed and you are not aware of it, or he is perhaps using but you are not aware of it, and you are not assisting him, that doesn't make you complicit. But from the very outset anyone in that house becomes a focus possibly of a crime.*
>
> (CEO of family support and advocacy service)

> *In some cases the police will arrive and start talking about a 'suspicious death' (rather than unexplained) and you can imagine the impact on the family of the kind of language used – and will then treat it as unexplained and therefore possibly a crime.*
>
> (Detective inspector)

Lack of explanation of the necessity for forensic examination and evidence gathering meant the bereaved lacked understanding of the work of officials:

> *It's always very difficult for families to understand why the police go in a bit like a bulldozer and then the family are left to try and pick up what's left.*
>
> (Coroners' service manager)

As with cot deaths and other unexplained deaths within the home, family members in shock at hearing news of the death can become alarmed to realise – or at least feel – that they themselves are under police investigation, assumed potentially guilty until the police satisfy themselves otherwise. This in turn threatens to undermine trust in the other authorities of the state to be encountered in the ensuing days, weeks and months.

Subsequent to this initial stage, professionals noted that the bereaved could experience further difficulties relating to the investigation of the death. A Scottish legal professional acknowledged inherent delays in the normal course of investigations that could leave families in limbo:

> Police officers will do their best at the time on the day. It's then reported into the Fiscal . . . But everybody is then kind of in limbo until we get the toxicology, maybe twelve weeks later . . . in the meantime these families are sitting in limbo . . . the post mortem has happened, the funeral is over and done with. But these families are still left sitting with a . . . that says 'death unascertained'.

A minister of religion noted the anguish and powerlessness that families could feel when in criminal cases defendants request a further post mortem:

> One of the things that has caused people a lot of anguish is that if there are charges to be brought, the realisation that the defence solicitor can . . . request another post-mortem, when families hear that the sense of powerlessness they feel, that their own feelings don't count . . . That state of powerlessness and the sense of anguish is really enhanced when they learn (that) if there are five people who can be charged, each one of them can have their own post-mortem.

In sum, the state's need to ascertain the cause of death, combined with the input of a range of specialist health and voluntary services, add to the complexity of the already difficult institutional terrain that any bereaved person has to traverse in the days and weeks after death. Few, if any, officials and practitioners are aware of the existence, let alone the functions, of all the other services and organisations, and rarely is one single person on hand to guide the bereaved through this terrain. Some required procedures, particularly those involved in gathering evidence to ascertain the cause of death, can disturb family members and undermine any confidence in dealing with the institutions that have to deal with bodies and survivors.

On contemplating the deceased's prospects after death, Shakespeare's *Hamlet* (Act 3, Scene 1) muses on

> the dread of something after death, The undiscovered country from whose bourn No traveller returns.

After a substance-related death, it is not only the deceased, but also those who mourn, who face an undiscovered country, a country full of unwelcome

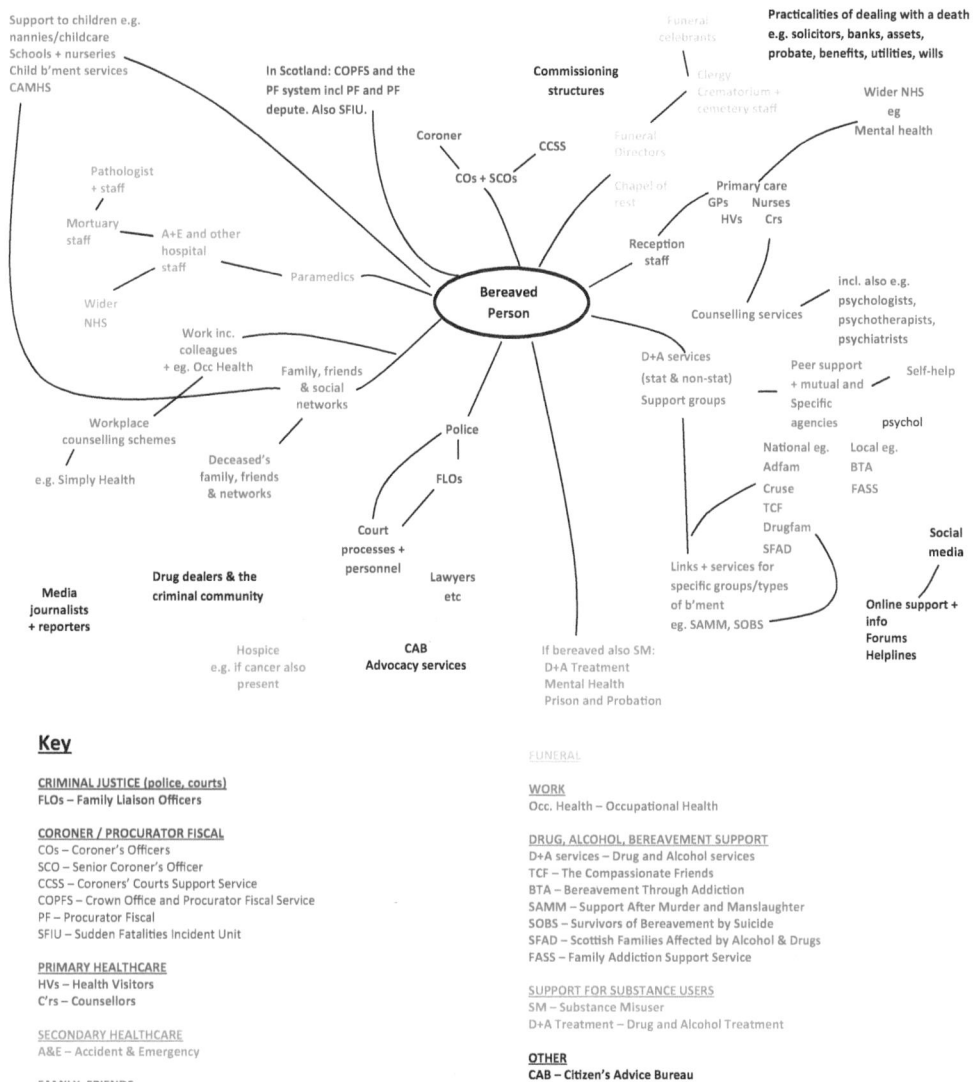

Support to children e.g.
nannies/childcare
Schools + nurseries
Child b'ment services
CAMHS

In Scotland: COPFS and the
PF system incl PF and PF
depute. Also SFIU.

Commissioning
structures

Funeral
celebrants

Practicalities of dealing with a death
e.g. solicitors, banks, assets,
probate, benefits, utilities, wills

Clergy
Crematorium +
cemetery staff

Wider NHS
eg
Mental health

Pathologist
+ staff

Coroner

CCSS

COs + SCOs

Funeral
Directors

Chapel of
rest

Primary care
GPs   Nurses
HVs   Crs

Mortuary
staff

A+E and other
hospital
staff

Paramedics

Reception
staff

incl. also e.g.
psychologists,
psychotherapists,
psychiatrists

Wider

NHS

Work inc.
colleagues
+ eg. Occ Health

Family, friends
& social
networks

Bereaved
Person

Counselling services

D+A services
(stat & non-stat)

Support groups

Peer support
+ mutual and

Self-help

Specific
agencies

psychol

Workplace
counselling schemes

Deceased's
family, friends
& networks

Police

FLOs

National eg.
Adfam
Cruse
TCF
Drugfam
SFAD

Local eg.
BTA
FASS

e.g. Simply Health

Social
media

Media
journalists
+ reporters

Drug dealers & the
criminal community

Court
processes +
personnel

Lawyers
etc

Links + services for
specific groups/types
of b'ment
eg. SAMM, SOBS

Online support +
info
Forums
Helplines

Hospice
e.g. if cancer also
present

CAB
Advocacy services

If bereaved also SM:
D+A Treatment
Mental Health
Prison and Probation

Key

CRIMINAL JUSTICE (police, courts)
FLOs – Family Liaison Officers

CORONER / PROCURATOR FISCAL
COs – Coroner's Officers
SCO – Senior Coroner's Officer
CCSS – Coroners' Courts Support Service
COPFS – Crown Office and Procurator Fiscal Service
PF – Procurator Fiscal
SFIU – Sudden Fatalities Incident Unit

PRIMARY HEALTHCARE
HVs – Health Visitors
C'rs – Counsellors

SECONDARY HEALTHCARE
A&E – Accident & Emergency

FAMILY, FRIENDS
CAMHS – Child and Adolescent Mental Health Services

FUNERAL

WORK
Occ. Health – Occupational Health

DRUG, ALCOHOL, BEREAVEMENT SUPPORT
D+A services – Drug and Alcohol services
TCF – The Compassionate Friends
BTA – Bereavement Through Addiction
SAMM – Support After Murder and Manslaughter
SOBS – Survivors of Bereavement by Suicide
SFAD – Scottish Families Affected by Alcohol & Drugs
FASS – Family Addiction Support Service

SUPPORT FOR SUBSTANCE USERS
SM – Substance Misuser
D+A Treatment – Drug and Alcohol Treatment

OTHER
CAB – Citizen's Advice Bureau

Figure 6.1 The terrain facing family members after a substance-related death

emotions and strange institutions. And even though many have travelled this institutional terrain before, few return to assist new arrivals – a rare exception being DrugFAM's helpline for those recently bereaved from substance use.[2] And even though practitioners and officials occupy this terrain, few take responsibility to inform and prepare new arrivals.

Challenges

A system that is fragmented and confusing for bereaved families presents several challenges to the practitioners and officials who work within it. The focus groups highlighted several factors that can affect practitioners' ability to respond appropriately to those bereaved by a substance use death. This includes the complex and continuously changing landscape of services challenges practitioners to keep knowledge and links to relevant sources of support up to date. Another issue is the influence of the workplace – its core functions, procedures, structures, underpinning policies, legislation and occupational culture. Pressures and priorities associated with the workplace were often perceived as barriers to communicating and building rapport with the bereaved. Focus groups also drew attention to the individual capability of practitioners, pointing to a lack of training in how to respond to bereaved people and/or substance use and addiction. Focus groups debated whether personnel can be taught to be compassionate or whether only individuals demonstrating empathetic qualities should be employed in certain roles. Although support groups have a valuable role, their composition can be particularly tricky after a substance use death. This second part of the chapter examines these four challenging aspects of the professional context: links with other professionals, groups and organisations; workplace priorities and culture; professional capability and training; and support groups.

a) Linking to other support

Supporting the bereaved depends on liaising with and referrals to other professionals, groups and organisations; our focus groups were troubled by how difficult this can often be. Joined up working is not commonplace, so communication and information sharing tends to be poor. A clinical psychologist working in substance use treatment reported variability in which organisations are or are not informed of clients' deaths.

> [W]hen we work with a client we will often be connected with other agencies and then maybe they die of a drug-related death and the information goes to some organisations but not to others so actually we could have been working with the person quite closely and we don't know until later that they've died – and the family expect us to know and we don't.

As a consequence, professional knowledge of the support available to the bereaved is often limited and depends on chance. The psychologist continued:

> It's still challenging in terms of knowing what services there are in your area . . . some-times it's coincidental that you know about a service – it can be very hit and miss and possibly there's duplication.

Also challenging is finding the time to keep up to date with a changeable service landscape on top of day to day workloads. This is particularly difficult for those professionals whose primary role concerns the deceased, not supporting the bereaved. A coroner's officer noted that:

> One of the things that's difficult for us is there's a lot of groups that have developed over the last few years and things have changed quite significantly in the number of support groups there are. But we're not aware . . . and unfortunately we haven't got the time to research, and if we were made aware we would be putting people in touch with support groups far more frequently.

Suspicion and lack of trust between professional groups were also believed to impede joined-up support for the bereaved. A funeral director (who works for a commercial enterprise) felt frustrated that her ability to support families was compromised by mortuary staff (who work for the National Health Service) withholding information on causes of death on grounds of confidentiality.

> I came through NHS and charity so I'm maybe not typical, but to me there's a huge role that we could take on if people gave us respect that I think many of us deserve and be actually let in to how people have died because that's not our right to know, so we don't often get to know how people have died unless the family have told us . . . I've always been bound by confidential-ity, it's something that I do, you know, it's sort of second nature. And now I'm in funeral directing I find this is a stick that's used to beat us with. You don't sign anything to say confidential so we can't tell you anything. So the NHS will never tell what somebody died of. They'll tell us "Oh you can't embalm that one. Don't touch it". But we're not told why because evidently we're not trustworthy enough so we are dealing with this person, we are advocates for the family, that's what we should be, that's what I am. And we are custodian of that body and to come from a place where I was trusted to uphold confidentiality into somewhere where now evidently I'm not, I find deeply frustrating.

If inter-agency working can be problematic, the following sections exam-ine factors within individual agencies that act as barriers to supporting those bereaved by a drug- or alcohol-related death.

b) Workplace priorities and culture

Many of the practitioners who may come in contact with the bereaved will not necessarily see support as part of their role; this has implications for how bereaved people experience services. Key practitioners here include police officers, paramedics, mortuary staff, representatives of the procurator fiscal service (Scotland) and coroner's service (England), although there will also be others. In the aftermath of a substance-related death, the bereaved will encounter professionals whose core functions concern the death and the dead rather than bereavement. Police officers are tasked with gathering evidence to investigate unexplained deaths that could be the result of a crime; mortuary staff's chief role is to look after the body, but they also have to facilitate viewing by relatives; and legal representatives will oversee processes that ascertain the cause of death. In such contexts, focus group participants indicated that professionals must focus upon fulfilling their primary duties.

> *If the police attend an unexplained death there is a level of investigation which has to be sort of carried out.*
>
> (Police drugs coordinator)

> *It's about understanding what that professional is there to do — and in the case of the police it's not their core function — it may be a part of their function. So when you've got other things to consider you and other things to do then.*
>
> (Police drugs strategy officer)

More specifically, a legal professional from Scotland suggested that preserving evidence on the body will take precedence over relatives' needs to access the body.

> *I've never had objections to families who feel that they want to touch and give a wee cuddle. The only time that they do it is there, it's the preservation of evidence they say no, and again that's there maybe they are thinking that there is going to be a charge for some reason, they think that perhaps there is still perhaps evidence on the body that [they] will get at the post-mortem because in general families get to view before the post-mortem. So at times you have got to be careful about preserving the evidence.*

Duties are informed and underpinned by legislation and organisational policies which can present challenges in responding compassionately to the bereaved. A senior coroner's officer explained that legislation dictates how police officers must proceed after an unexplained death.

> *[T]he Police and Criminal Evidence Act is so stringent that they have to follow set rules so that if they got to court and they did want to prosecute the supplier, the dealer or something like that and there was one thing that hadn't been followed the whole thing would have been thrown out. And then you see the headlines, you know*

> *'Dealer gets off again because police didn't do their job'* . . . *everybody does have their sort of role to play.*

Professionals and officials who wish to provide greater support to people bereaved through substance use can find such rules frustrating. One difficult issue is access for families and others to the body once removed for post-mortem. As well as access being restricted in order to preserve evidence, deaths involving class A drugs can be deemed to present a risk of infection, which – depending on local policy – can affect viewing of the body. A mortuary services manager noted, with regret, that this prevented family members or others from being close to, or touching the body.

> *When we have any drug related deaths come into the mortuaries, they are, if it is class A drugs they are immediately classified as potential risk of infection until we are advised otherwise by medical staff. And this has a wee bit of a knock on effect unfortunately in that the policy within and as far as I am aware it's just restricted to [our health board] the policy is that any family viewing . . . is behind a glass screen.*

Neither she nor her colleagues were satisfied with this policy.

> *I think the majority of staff that work certainly in the mortuary are of the opinion that that policy is not right and that we are not offering the best support to bereaved relatives. However you know that's the policy so we do feel that we are letting relatives down by not allowing them to be with their loved one.*

Another professional whose service supports families affected by substance use and operates in the same health board area, recalled a mother who pleaded with mortuary staff to let her be with her son.

> *It was her only child as well, she had gone to, it was the [hospital] and she had pleaded with them because it was the screen that, and they eventually let her hold his hand, and the difference that made.*
>
> (Manager of family support service)

A coroner's officer spoke of the resource implications of recent legislation requiring most inquests to be held within six months of the death.

> *What's happening now, the new Coroners and Justice Act and with the new Chief Coroner in place, there is a drive now for all inquests, where possible, to be heard within 6 months . . . there now is huge pressure on the coroners' service to act without additional resources to get those inquests heard within six months . . . The resources for the coroner's service are really not sufficient for the work that we're doing. I have noticed that in the twenty odd years that I've been doing the role, my face-to-face*

contact with families has diminished immensely. And I think to the detriment of our service to people who are bereaved, but also to the detriment of me as a coroner's officer doing my job because I am frustrated because I want to give that more personal service.
(Senior coroner's officer)

A mortuary services manager also spoke of resource constraints.

I think as well staffing issues in the NHS mean that we are not allowed to spend as much time with grieving relatives as we would like.

How substance use treatment services are structured was also experienced as a barrier to supporting bereaved family members. Treatment services are designed to serve people with substance use problems and do not routinely support those around them who may be affected. Substance treatment services may have nothing to offer bereaved relatives. One treatment professional was told by her employer that if she wanted to visit a bereaved mother, she would have to do so in her own time.

She would come into the drug service, the mother, the grandmother, the woman who had lost all her children . . . Because she had no one else to speak to . . . So you know we are talking about extraordinarily isolated people with very small groups of relationships who you know and so it's kind of that thing of, the importance of addiction services, that we often should and can hold for parents and it's not neces-sarily within the organisational culture of addiction services to be doing that. So for example I remember being told that I would need to go around and see this woman in my own time.
(University teacher with background in drug treatment and support)

Workplace culture also influences how professionals are able to respond to substance-related deaths. Sometimes one occupational group is frustrated by the culture of another.

I work in a drug and alcohol service and if one of our service users die, our first point of contact is often the police. And not all the time but sometimes it has been very good, but sometimes you get this, you get that expectation . . . I've even had a police officer say to me over the phone, "Oh well it was expected, it's not a surprise is it?", which just isn't helpful you know.
(Focus group 2, manager of family support service)

A police officer in the same focus group corroborated this view.

Unfortunately there are so many unexplained deaths that are . . . related to drug use, consumption that you know for me personally it's almost become, it's accepted isn't it?

A freelance journalist provided insight into the culture of newspaper offices where time pressures, particularly for the national press, require staff to produce articles in a matter of minutes with possible consequences for portraying the person who has died with sensitivity.

> *Local newspapers may tend to be more sensitive to the individual. But national newspapers have 5 minutes to do an 80 word review so can be more brutal.*

In sum, some workplaces do not, and some cannot, prioritise support for bereaved families. Workplaces vary in the extent to which, as organisations or as individual practitioners, they have the desire or resources to go the extra mile for families.

c) Professional capability and training

Our focus groups considered professionals' personal capabilities with regard to skills, qualities, knowledge, attitudes and experience of people bereaved through substance use. Focus group members felt that professionals who are ill equipped in these areas can struggle to respond with skill and compassion. Referring to police attending the scene of a death, a professional working in family support suggested:

> *It shouldn't be hit or miss, it should be somebody who is more of a people person reporting to the scene.*

A Church of Scotland minister emphasised the importance of empathy – the refusal to 'other' the bereaved family discussed in Chapter 3.

> *It's not just a going in . . . and I'll just say that at the funeral service and that's it all. It's a continuing relationship and then it's coming back about being a human being and non-judgmental, a human being just there as a warm engaging presence, as . . . who thinks, you know, what if this was my son or my daughter?*

Focus group members felt that empathy needs to be supported by knowledge of substance use and addiction.

> *Some understanding of drug and alcohol use helps in terms of your attitude and how you bring yourself to deal with someone. The recognition that people are likely to feel stigmatised and I think that whole thing about shame and how when we feel ashamed we withdraw and at the very least we drop our eyes and look away, and actually I won't even go into the surface. I think just to know that that's where some- one could well be at and therefore you've got to overcome that hurdle to meet them.*
>
> (Counsellor/trainer)

Knowledge is more effective when complemented by experience. A Scottish police officer noted that uniformed officers attending a death may be relatively inexperienced in handling the situation.

> *Largely in the first instance it is going to be uniformed officers which potentially might be sort of relatively inexperienced that will be there in the first instance.*

Families may encounter inexperience in other occupational groups. The local journalist attending an inquest could be in his or her first job. An English focus group member who had lost her son to a drug overdose found that an advocacy worker employed to represent her at the inquest had no previous experience of cases involving substance use.

> *[T]he person who's taken on my case . . . he did turn round to me at one of our sessions and say to me "How experienced do you think I am with drugs then?" and I said "I don't know, I just presume you are", and he said "This is my first case". Now that just, to me, says it all. My son's case is the first one he's ever taken up as an advocate. He's going to speak for me at the inquest.*

Drug and alcohol workers' lack of experience and expertise in dealing with bereavement might, it was suggested, cause them to resist adding bereavement support to their role.

> *I think it's something we probably don't all think of . . . who is responsible? I mean, you know, from who is working with the client and the client dies, should they con-tinue some support for the parents if the parents want it. Or do they feel that they want to, because I think that some people would probably feel that they wouldn't be competent enough to deal with that etc. so there might be issues to do with that. They may feel that it may not be part of their job to deal with bereavement because of the bereavement services out there.*
>
> (Alcohol and drug partnership worker)

Experience, however, needs to be complemented by expertise and discernment, treating each person as an individual.

> *It's also stepping away from your own sense of judgement even if you see 20,000 deaths through drug and alcohol use it's got to be for each person.*
>
> (Bereavement counsellor)

Despite highlighting shortcomings in professional skill sets and attributes, focus groups pointed to a range of professionals who lack training in responding to substance use and addiction. Police officers, general practitioners and social workers were mentioned in this regard.

Social workers get shockingly small amount of compulsory drug and alcohol [training], it's ridiculous.

(Policy director for family support service)

The police don't have an awful lot of training related to drugs and alcohol at all any aspect of it.

(CEO of family support service)

A representative of a Scottish bereavement support organisation admitted that, although its workers are experienced in responding to alcohol-related bereavement, they need more training in order to respond adequately to drug-related bereavement.

I suspect that we probably see more people, relatives of alcohol related and yes some drug ones, possibly it is a minority, and to increase our involvement with relatives of drug users, I think might require a bit of organisation and more training.

Focus group members differed over whether it is possible to train professionals to respond appropriately to the bereaved, especially with regard to compassion. Several thought not – for them, compassion is, or should be, innate.

I think there are people out there who innately know what to say and actually it's not about the clever things you say, it's not about that phrase that you always have when you answer the phone, it's about picking up the phone and having somebody on the end who's distressed and just saying "I'm really sorry, that's awful. Talk to me". And I don't think that that can be taught.

(Funeral director)

This prompted discussion of whether those without such innate qualities and skills should be employed in jobs that serve the public.

People should not be in community service like my service if they can't communicate properly, and if they've got barriers to their communication.

(Senior coroner's officer)

The opposing opinion, however, that it is possible for professionals to develop the appropriate skills and qualities, was evidenced by the widespread rollout of child protection training.

For child protection everybody has to be trained at a level suitable to their job role and that's built in to their job descriptions now as well. I think what's coming up out of this is this is a specialism, this is a very important area that we actually need to start addressing.

(Senior support worker)

Leadership was identified as key to bringing about cultural change within workplaces and enhancing professionals' skills.

> *If leadership sets the tone then it does have a great potential to dissipate its way through. I think if the tone is you know, I think a warm engaging human person can still be ultra-professional in response to an investigation, absolutely.*

(Ex-police officer)

The possible emotional strain on professionals of working with people bereaved through substance use (Chapter 3) was also discussed. Supervision for this was considered to be variable. Although not specific to substance use bereavement, a representative of an English bereavement support organisation gave the example of some NHS workers to whom she had delivered training.

> *I did some training for NHS staff on bereavement and, though they were working with dying people, none of them had the opportunity to unload at all.*

In sum, all focus groups identified personal capabilities regarding substance use bereavement as an issue; however, not all participants agreed on how to remedy this or what kind of training would be most effective for professionals and officials who encounter this kind of bereavement from time to time but not every day.

d) Support groups

As noted earlier in our discussion of the 'system's' fragmentation, focus group members were very aware of the needs of people bereaved through substance use being complex and diverse, so support has to be sensitive and nuanced – certainly not 'one size fits all'. Here we consider this in relation to support groups.

Some focus group members considered that support groups could help people bereaved through substance use. They observed, however, that most groups are for those affected by another's current substance use rather than dedicated bereavement support groups. They recognised that the group most suitable to any one individual may not be obvious. Some bereaved individuals may find it difficult to attend a group where some or most members' close others are still alive and engaged in substance use. Meanwhile the bereaved may represent an unspoken fear for those members with close others still living.

> *I think a lot of the family support groups that exist around drugs and alcohol, although some of them do have expertise in bereavement ... some of the others don't and we've had people reporting that if they go into the group and they're bereaved and everyone is struggling with a relative who's got a very current problem, actually they don't feel comfortable because they don't want to share that because it's a different paradigm of experience.*

(Policy director for support service for families affected by substance use)

Participants perceived that instead the bereaved would find more and better support amongst those who were similarly bereaved.

> *So quite often maybe the only support they can get is through other people who have had that experience.*
>
> (Procurator fiscal)

However, friendships established in a family support group while a close other was alive would then be lost if that person died.

> *It's important that there are separate groups for those bereaved, but then as you say you take away their support group and friends in that group environment.*
>
> (Senior coroner's officer)

Participants were aware that some bereaved people may not want to access support that is in any way associated with substance use.

> *Any bereavement may be complex for a variety of reasons, so for some people generic services like Cruse might be more palatable if they don't want to address the drugs aspect.*
>
> (Clinical psychologist)

This was seen to be particularly pertinent to people who had lost close others to substance use where this was not a regular pattern of behaviour and there was no previous identification with treatment providers or family support groups. In this regard, participants were concerned that the bereaved may isolate themselves from those best placed to help them because of pre-conceived perceptions of what constitutes a substance-related death.

> *I am just thinking of two situations recently where young people have died from a drug overdose and in fact the families when they got the letter didn't know what the drugs were and called back to me and said what are these? And I think that is particularly true given the massive number of drugs that are now around legally or illegally or partially or whatever and transitionally, and so wouldn't identify themselves as having necessarily a child who had addiction. And so there is, and so they do in fact have a child who has died from a drug related overdose. So there seems to be a kind of bubble there or a gap there, or a something there where they certainly wouldn't come to family support services for addiction.*
>
> (Procurator fiscal)

Research on mutual help support groups indicates that they 'fuse' together people who feel they share a similar experience, but persons who sense their experience to be different from the group norm will feel uncomfortable and be unlikely to return; they then either find another group, find some other kind of support or abandon their search for support (Rock, 1998). The experience of losing someone close to drugs or alcohol can be particularly isolating, so there needs to be considerable flexibility of provision. Some may feel supported by a general bereavement group, some by a group for those affected by a living person's substance use that they already belong to, some by a specialist group for people bereaved by substance use. Others may prefer counselling, whether generic or specialist bereavement counselling. Given what we have already seen in the first part of this chapter about 'the system's' fragmentation and complexity, getting information about available support disseminated to all the relevant officials with whom the bereaved may come into contact is clearly challenging. It is also challenging for practitioners to know what kind of support might be most appropriate for a particular client, and of course the waiting list for the most appropriate support may be longer than for less appropriate support.

Conclusion

Based largely on our focus groups, this chapter has analysed 'the system' that comes into play after a drug- or alcohol-related death. The first part showed how, far from being clear or joined up, this 'system' is in effect a very complicated undiscovered country for bereaved family members, a country for which this chapter offers possibly the first ever map (Figure 6.1). We hope that officials and professionals can simplify and personalise this map in ways that help inform the bereaved family members with whom they come into contact; it may also help practitioners themselves to understand other organisations – their duties, the limits to their duties and the pressures and challenges they face.

Our focus groups contained a wealth of awareness and understanding of substance use bereavement; their unique perspective allowed them to identify a number of challenges that may impede even the most aware and knowledgeable practitioners from responding appropriately to the needs of the bereaved. First, participants argued that supporting the bereaved is made more difficult by an ever-changing service landscape where information sharing and joined up working between organisations is not the norm and in some instances is undermined by a lack of trust between services. Second, there is the workplace itself where organisational structures, core functions and culture, both separately and together, can limit sensitive responses. Third, personal capability and access to training on substance use and/or bereavement can be a barrier to effective care, although some suggested that the necessary skills and qualities, particularly compassion, cannot be taught. And fourth, the particularly isolating experience of substance use bereavement makes it challenging to match support group and individual. These findings are consistent with the literature on inter-professional working, which has identified how health, social care and children's services may be impeded by differing work cultures, lack of understanding of each other's roles and responsibilities, pressures of time, limited resources, competing priorities and insufficient access to other professionals (see, e.g., Bailey et al., 2006; Braithwaite et al., 2012; Larkin and Callaghan, 2005; Snelgrove and Hughes, 2000).

The challenges for practitioners in responding adequately to those bereaved by substance use are so substantial and widespread that significant change in this area is likely to be protracted and incremental. Yet our focus groups showed that the complexities of substance related bereavement are not completely unknown to practitioners, and very well understood by some; this provides a strong foundation for raising awareness and understanding amongst other practitioners. This is the main purpose of the practice guidelines which we will introduce in the next chapter.

Notes

1   www.gov.uk/after-a-death/organisations-you-need-to-contact-and-tell-us-once
2   www.drugfam.co.uk

References

Atkinson, M., Jones, M., and Lamont, E. (2007). *Multi-Agency Working and Its Implications for Practice: A Review of the Literature*. Reading: CfBT Education Trust.

Bailey, P., Jones, L., and Way, D. (2006). Family physician/nurse practitioner: Stories of collaboration. *Journal of Advanced Nursing*, 53(4), 381–391.

Biddle, L. (2003). Public hazards or private tragedies? An exploratory study of the effect of coroners' procedures on those bereaved by suicide. *Social Science and Medicine*, 56(5), 1033–1045.

Braithwaite, J., Westbrook, M., Nugus, P., Greenfield, D., Travaglia, J., Runciman, W., Foxwell, A. R., Boyce, R. A., Devinney, T., and Westbrook, J. (2012). A four-year, systems-wide intervention promoting interprofessional collaboration. *BMC Health Services Research*, 12(99), 1–8. DOI: 10.1186/1472-6963-12-99

Chapple, A., Ziebland, S., Simkin, S., and Hawton, K. (2013). How people bereaved by suicide perceive newspaper reporting: Qualitative study. *British Journal of Psychiatry*, 203(3), 228–232.

Kastenbaum, R. (2007). *Death, Society and Human Experience*. 9th Edition. Boston: Pearson.

Larkin, C., and Callaghan, P. (2005). Professionals' perceptions of interprofessional working in community mental health teams. *Journal of Interprofessional Care*, 19(4), 338–356.

National Council for Palliative Care. (2014). *Life After Death: Six Steps to Improve Support in Bereavement*. London: The National Council for Palliative Care.

Peters, K., Cunningham, C., Murphy, G., and Jackson, D. (2016). Helpful and unhelpful responses after suicide: Experiences of bereaved family members. *International Journal of Mental Health Nursing*, 25(5), 418–425.

Rock, P. (1998). *After Homicide: Practical and Political Responses to Bereavement*. Oxford: Clarendon.

Snelgrove, S., and Hughes, D. (2000). Interprofessional relations between doctors and nurses: Perspectives from South Wales. *Journal of Advanced Nursing*, 31(3), 661–667.

Templeton, L., Ford, A., McKell, J., Valentine, C., Walter, T., Velleman, R., Bauld, L., Hay, G., and Hollywood, J. (2016). Bereavement through substance use: Findings from an interview study with adults in England and Scotland. *Addiction Research & Theory*, 24(5), 341–354.

Valentine, C., and Bauld, L. (2016). Marginalised deaths and policy. In Foster, L. and Woodthorpe, K. (Eds.). *Death and Social Policy in Challenging Times*. Basingstoke, New York: Palgrave Macmillan, Ch. 7: 110–128.

Valentine, C., Bauld, L., and Walter, T. (2016). Bereavement following substance misuse: A disenfranchised grief. *Omega Journal of Death Studies*, 72(4), 283–301.

Walter, T., Ford, A., Templeton, L., Valentine, C., and Velleman, R. (2015). Compassion or stigma? How adults bereaved by alcohol or drugs experience services. *Health and Social Care in the Community*. doi 10.1111/hsc.12273

Chapter 7

Improving the response of services

Peter Cartwright, Lorna Templeton and Gordon Hay

Introduction

This chapter describes our 'pathway to impact' – our strategy for ensuring that our research findings make a difference to how people bereaved by substance use are responded to and supported by the practitioners they encounter. This strategy comprised developing and disseminating a set of good practice guidelines, which were produced by the research team and a diverse range of practitioners and officials. First, we consider the need for such guidance. Second, we describe how the research data and ideas from wider policy research were drawn together to identify the five key messages that formed the backbone of this guidance. Third we describe how the guidelines were produced by a working group of experienced and diverse professionals and workers and reviewed by an ad hoc group of other experienced people. Fourth we discuss how the guidelines have been disseminated and used within the UK to date. We conclude with some reflections on this important final stage of this study.

The chapter is written by three of those involved in producing the guidelines. The guidelines may be accessed at http://rebrand.ly/bereavementguidelines.

The need for good practice guidelines

It was important that the research had a practical application. This was achieved through developing good practice guidelines that would be accessible to a wide range of people. Three main factors drove the need for this guidance:

First, as earlier chapters have shown, these are typically severe and complicated bereavements, producing in bereaved people a high need of support but who, due to stigma, are often reluctant to engage with professional support.

Second, the guidelines would be the first of their kind, filling a much-needed gap in supporting professionals and other workers to respond to a group of bereaved people who have been largely ignored by policy and practice. The research findings from both bereaved people and the focus groups painted a mixed picture in terms of response: some good, some inadvertently poor and some prejudiced (Templeton et al., 2016; Chapters 2 and 6 this volume). This reflects growing concern in a number of countries about a 'compassion deficit' in health and social care (Ballatt and Campling, 2011; Cole-King and Gilbert, 2011; and Walter et al., 2015).

Third, the guidance would be evidence based, informed by both the wider policy literature and by the interviews and focus groups with 146 bereaved adults and professionals. In addition, it would draw on the working group members' first-hand experience of working with people bereaved through a substance related death.

In considering the target audience for the guidelines the research team turned again for guidance to the interview and focus group findings. These showed that adults bereaved by substance use can interact with a wide range of individuals, services, organisations, networks and others, all of whom could in some way offer support (See Chapter 6, Figure 6.1). Therefore, it was considered essential that the guidelines be *generic*, potentially relevant to anyone on that support map, as well as any others not included but whose work still brought them into contact with people bereaved by substance use.

Identifying the five key messages

This section outlines the development of five key messages that form the backbone of the guidelines and describes the other preparatory work that informed the working group's task of producing the guidelines. It summarises first the ideas taken from the wider policy literature and then ideas from the ongoing analysis of the research interview and focus group data.

The wider policy literature

This task involved a scoping review of policy literature and other resources and guidance in three main areas: professionals who come into contact with bereaved people, bereaved people themselves and services and organisations who deliver support to bereaved people. The wider bereavement literature was considered, identifying resources specific to substance use or other 'difficult deaths'. We also found some potentially useful ideas in the literature on supporting those at the end of life and their families/carers.

The overriding finding from this scoping review, which confirmed what the research interviewees and focus group participants had reported, was that there was a major gap in specific guidance for professional and worker groups who may come into contact with those bereaved by substance use. The exception is the small number of specialist resources and services which have been developed for bereaved people themselves, summarised in the guidelines (Cartwright, 2015). Overall, there is a wealth of information for responding to bereavement more generally, and we found some ideas here which resonated with the key themes and messages emerging from our research (see Figure 7.1 for some examples) and which informed the guidelines. However, although many of the

"We know that the manner in which services, professionals and volunteers respond to those who are bereaved can have a long term impact on how they grieve, their health and their memories of the individual who has died". (Cruse Bereavement Care[1] and the Bereavement Services Association, 2014)

"Please be aware that in most legal situations a person who had died is referred to as the deceased. This convention has been used in this booklet. Coroners and their staff understand that the person who has died was a unique individual". (Ministry of Justice, 2014)

"Please note that we never refer to 'committing suicide'; this expression has remained in public usage long after suicide legally ceased to be considered a criminal act. Its use can distress families and it is preferable to refer to 'death by suicide' or that the deceased 'took their own life' or 'ended their life.'" (Winston's Wish[2] – quote taken from their website, accessed April 1, 2016)

Figure 7.1 Examples of messages from the scoping review

principles of supporting bereaved people are applicable and transferable to this population, the scoping review also confirmed the need for specific guidance to offer insight into this group's complex, nuanced and diverse experiences.

Ideas from the interview and focus group studies

Given that the guidelines would focus on *responding* to this group of bereaved people, the analysis of the interview and focus group data focused on themes related to *support* and interviewees' responses to being asked what advice they would give to others who had experienced the same type of bereavement. As a result we identified eight emerging themes from the interviews, and seven messages from bereaved interviewees (Tables 7.1 and 7.2). This information was circulated to all Working Group members prior to the Group's first meeting.

Emerging themes from the interview analysis

From an analysis of bereaved peoples' experiences of accessing or receiving support from a range of sources we identified eight main themes (Table 7.1) to inform the task of the working group. Each theme was illustrated with two quotes, one showing a helpful or supportive response, the other a less helpful or unsupportive response. It is not possible to describe all eight themes, but they have been grouped in pairs, each pair being illustrated with one positive and one negative example (also see Chapters 2 and 6).

Keeping the family/close others informed, and being available and taking time

> *A person who helped us absolutely incredibly at the most crap time was [my son's] doctor ... he came round and he was absolutely incredible. He was so understanding and we talked ... that was the best counselling that we had.*

(MotherE)

Table 7.1 Support themes

1. Continuity of/gaps in support
2. Keeping the family/close others informed
3. Showing/not showing kindness, concern and tact
4. Respecting/not respecting privacy
5. Being available and taking time
6. Media responses
7. Stigma
8. Generic or specific support, and meeting others with similar experiences

Table 7.2 Messages from bereaved interviewees

Message	Example
Be kind to yourself	"I would tell them that it's important to find time to be with how you feel – because the fear of feeling your feelings leads to a worse outcome than facing them" (Father talking about son)
Do not blame yourself	"[W]hen they start thinking 'what did I do wrong?' . . . that thing 'what did I do wrong?' but you didn't, you done the best . . . they do the best they can and then things happen" (Husband talking about wife)
It gets easier with time	"I would like to say, 'you'll be okay . . . Life will – you know, you will get out of this.' I think I never ever thought I would ever get over it . . . The actual physical pain as well, I never thought I would get over it . . . I thought how could I possibly live a normal life . . . how am I going to live a normal life with this massive thing that's happened to me" (Daughter talking about father)
Everybody deals with bereavement in their own way	"I would just, the only thing I would say to them is, deal with it in your own way, everybody's different like . . . Because I think my sister grieved a lot differently to me . . . Even though we were both in the sort of same situation, we both done it differently" (Son talking about father)
Understanding addiction can help	"Really alcoholism, it's a hell of a problem to deal with. And I think anybody trying to deal with it has to accept that there are limits on what you can do because ultimately each person chooses his own life. And I think that's an important message . . . Do what you can by all means but it may be you can't solve this problem. Maybe you can't crack it. It may be that with all your best efforts the problem will still be there and the problems may still get worse and in the end it may result in death" (Father talking about son)
Dealing with stigma	"I would say that they shouldn't, like I did, they shouldn't stigmatise themselves, they shouldn't put themselves down or worry about what other people are going to think" (Brother talking about brother)

(Continued)

Table 7.2 (Continued)

Message	Example
Consider seeking extra support	"I would probably say to them I think it would be a good idea for them to go and speak to somebody . . . To have counselling or something . . . and I would advise them to join a bereavement group as well . . . where other people are feeling the same and other people have the same loss because you do feel, it's one place you do belong" (Mother talking about son)

I did feel desperately let down because all I know the facts are that there was a young man fighting for his life and surely somebody in that A&E department would have had the decency to say, "Well I think he needs next of kin" but no, they didn't.

(MotherE)

Showing/not showing kindness, concern, tact and respecting/not respecting privacy

And although the police stayed in touch, but that was for their own reasons, because they needed these statements, the information at the inquest, but again it would have been nice if they'd said, Are you okay? Is there anything you need?

(MotherE)

They just said that they needed to talk to me about something, so I immediately said; oh it's [my partner] isn't it? . . . And they said, we do need to speak to you and it is urgent . . . they were . . . as gentle as they could be, I think they asked me was there somewhere where we could go to speak in private . . . I took them to one of the little bit of a restroom where we sit and have a cup of tea and they [told me].

(PartnerE)

Stigma and media responses

It was in the [local paper] . . . [the journalist at the inquest] was taking notes and she said, I'm going to report it anyway. Could we work together on this as opposed to me just writing the story up and disappearing back to the office? . . . She wrote it up and she emailed it to me and she said have a look. Is this okay? This will be the actual wording. Are you happy with it? Is there anything you want to change? So I changed a few things, and that was it.

(FatherE)

There is a big stigma, nobody knows what to say and they almost kind of look down on you as if it's some kind of lower class sort of problem that . . . we don't talk about

things like that. I think well actually we should talk about things like that. I think there just needs to be more . . . publicity around the impact that it has on other people . . . there's not much about actually what it does to your relationships and what it does for the people that are left behind.

(DaughterE)

Continuity of/gaps in support and generic/specific support and meeting others with similar experiences

And we went on like that, really, for a couple of years . . . that's when I discovered [organisation] . . . And that was a lifeline . . . Because all of a sudden I realised I wasn't alone anymore, or we weren't alone anymore . . . And that was just incredibly helpful. I mean, it didn't actually make any practical difference, but it made a lot of sort of emotional difference . . . because we could talk to people who understood how we felt.

(MotherE)

I went for an assessment [for counselling] . . . and I haven't heard anything since. . . [That was] about four months ago.

(MotherS)

Messages from bereaved interviewees to other bereaved people

The working group members received a list of themes summarising the advice that our bereaved interviewees would give to others who had also been bereaved through substance use. Each theme was illustrated by a few quotes. Table 7.1 lists these themes, and Table 7.2 gives examples of the quotes used to illustrate the themes.

Finally, drawing on this, we identified and then drafted five key messages for professionals and workers. These would inform the working group discussions and serve as the backbone of the guidelines (Figure 7.2). With all of this preparation completed, the working group was formed and the process of developing the guidelines begun. This process, and the evolution of the guidance for the five key messages, is outlined in the next section.

How the guidelines were developed

The research team decided that the best way to produce the guidelines was to form a working group of people who had first-hand experience of working with people bereaved through a substance-related death. The group comprised 12 members, representing services covering England, Wales and Scotland and a wide range of relevant professional and personal experiences: the senior policy and research officer from Alcohol Concern[3] with a particular interest in the

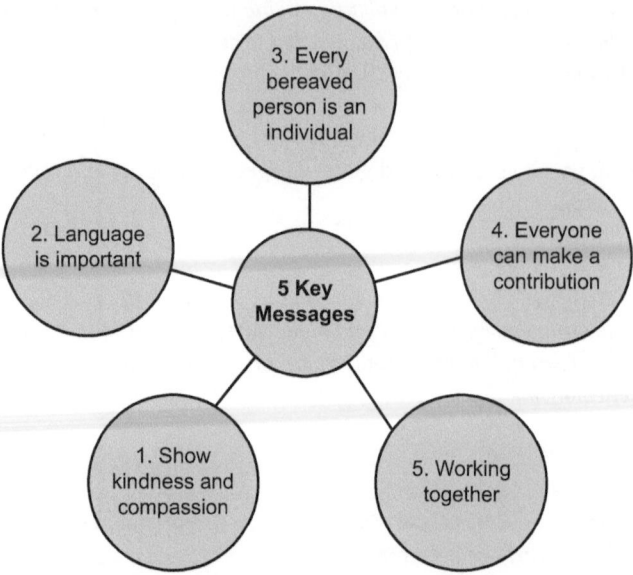

Figure 7.2 Bereavement guidelines: the five key messages

language used to describe people who use alcohol and drugs; a senior coroner's officer who was also involved in the training of coroners; a funeral director with experience of working with families bereaved through a substance-related death; a detective sergeant with family liaison experience such as informing people a relative had died through substance use; a NHS doctor working in general practice and interested in patients who use substances; an experienced paramedic involved in the training of paramedics; a university chaplain who had conducted many substance-related funerals; and those involved in bereavement counselling, supporting families affected by substance use and in drug/alcohol treatment. Some members had also been bereaved through a substance-related death or had a family member who had a current or past problem with substances. Selected members of the research team attended all the working group meetings and were closely involved in the process of writing the guidelines.

The working group was led by a chairperson who was independent of the research team and had experience both of facilitating groups and working with people bereaved through a substance related death. The chair's tasks were to facilitate the group to produce the guidelines, to be the primary author and to co-ordinate and assimilate the comments generated by the two reviews of draft versions of the guidelines. The chair reported to the research team in what was effectively a 'contractor' to 'client' relationship, because the working group was producing the guidelines for the team. Therefore, parameters were set for the group, such as basing the guidance around the five key messages and the project time frame. However, this relationship was also collaborative, with the research team trusting both the chair and the group to produce the guidelines while still monitoring their work and fully supporting them. This relationship was considered to have contributed to the success of producing the guidelines.

Particular attention was given to forming a working group that was cohesive, motivated and worked collaboratively and where individual members took ownership of the task and felt that their unique contribution was valued. This included fully using members' first-hand experience and the good practice ideas they had developed through their work. It was also important to trust members to work in the way they saw best and welcome all their contributions.

In summary, there was a successful balance between the negotiable and non-negotiable elements of the process, which created an environment of openness, trust and creativity, within set boundaries.

The research team wanted two drafts of the guidelines to be reviewed by a group of people representing wide ranging experience in the field. In addition to working group members, these included members of the project's advisory group (which oversaw the project and provided direction and guidance, as discussed in the Introduction); representatives from the national organisations Adfam,[4] the Adfam/Cruse BEAD project (Bereaved Through Alcohol and Drugs), DrugFAM,[5] Cruse Bereavement Care, Cruse Scotland,[6] Scottish Families Affected by Alcohol and Drugs;[7] a freelance journalist to ensure that the guidance was relevant to the media; and finally two people to advise on legal processes following a death in Scotland.

This review process generated a wealth of useful suggestions, particularly about specific guidance relevant to the five key messages, which ensured the guidelines met their aim of being helpful and appropriate to any professional or worker. The final document reflects the most helpful and appropriate of these suggestions, as keeping the guidelines concise did not allow them all to be used.

The guidelines were produced over a 28-week period, ending with their launch on June 23, 2015. Table 7.3 summarises the key stages of this process.

Engaging the reader

The chair, supported by working group members, gave considerable attention to who the likely readers of the guidelines would be and how best to *engage and motivate them* actually to use the guidance. He recognised that readers are probably very busy, have limited time and may have the normal human resistance to change. Even the most motivated reader was thought likely to be distracted by other demands.

In addition, potential readers were thought to lie across a spectrum of willingness to engage with and then use the guidance. At one end are those who already

Table 7.3 Development of the guidelines

Timeline	Activity	Comments
Week 1	Meeting between research team and working group (WG) chair	Consideration given to what aspects of the task were fixed, such as using the five key messages, and what were not, so where the group's creativity and ideas could be maximised.
	Chair emailed all WG members to introduce himself; to provide them with the background information described earlier in this chapter; and to ask members to begin thinking about what guidance was needed and how that should be communicated to the reader	This email effectively started the group process and introduced group email as the primary way the group would communicate.
Week 6	First WG meeting	To complete forming the group, detail the task and consider how the group would complete the task
Weeks 6 to 12	WG members produced ideas for the guidelines and considered how they needed to be written	WG members chose to work in regional sub-groups. These sub-groups used the background information and their experiences to produce good practice guidance about each of the five key messages. One member chose to work alone.
Week 12	Second WG meeting	Members' work was discussed and reviewed; agreement on the broad content of the guidelines; and on how they would be written.
Weeks 12 to 17	Chair wrote first draft, reviewed by research team and revised accordingly	
Week 17	First draft out to review	Sent to WG members; the research team; the project's advisory group; and a small number of additional experts to ensure input from national organisations supporting bereaved people and/ or people whose lives are affected by someone else's substance use.

Timeline	Activity	Comments
Week 19	Review completed	
Weeks 19 to 22	Writing second draft, initially by chair and then by chair and research team	
Week 22	Second draft out to review	Sent to the same people as first draft.
Week 24	Review completed	
Week 26	Final draft written by chair and research team	Followed by the production of a PDF and 2,000 hard copies.
Week 28	National launch of the Guidelines	The project's final event held in London.

appreciate the importance of supporting bereaved people and are motivated to develop their working practices, some of whom would probably not need these guidelines. At the other end are those who may have been told they have to read the guidelines by their manager but who lack motivation to engage with the guidance seriously. We therefore accepted there are limits to what any set of guidelines could achieve. However, in between these polarities we anticipated there to be a relatively large number of readers, so we focused on how best to engage them by:

- Keeping the guidelines generic so they were relevant to all professionals and workers, for example, using the term 'good practice' over 'best practice' to recognise the diversity of work settings, and that no single approach was necessarily the 'best'; not assuming the reader had specialist knowledge of either substance use or bereavement, whilst avoiding being patronising to those who did.
- Using a 'less is more' approach. Each key message was kept to two pages, thereby reducing the core guidance to just ten pages.
- Engaging readers by addressing them directly, inviting them to empathise with people bereaved through substance use by asking them questions about their own personal lives and about how they work.
- Treating the reader with respect, as a peer and equal, for example, making clear that the guidelines were written by the reader's peers; using 'you' and 'we' to address the reader; 'asking' rather than telling or directing; and ideas being 'suggested' for 'consideration'.
- Focusing on *specific practical* guidance to help the reader do their job and invite them to consider what they see as relevant to their work.
- Being consistent with the processes of all relevant UK institutions, including Scotland's specific legal processes.

- Offering all the basic information required to support a bereaved person, that is, the guidelines offer a 'one stop shop' containing all necessary basic information and guidance.

Overcoming stigma and shame

The guidelines encourage the reader to empathise with the likely stigma and shame felt by a bereaved person and they suggest practical ways to reduce the risk of provoking a bereaved person to feel stigmatised and ashamed. This was particularly relevant to key messages 1 to 3.

The high degree of stigma that people bereaved through a substance-related death often experience is clear from the research (Walter et al., 2015; Templeton et al., 2016; Chapter 3, this volume). Anyone who is stigmatised usually feels shame if they believe what the stigma implies (i.e., that they are somehow unacceptable or defective). The typical response to shame is a powerful urge to withdraw. Alternatively, they could feel angry if they do not believe what the stigma implies about them. Either way, a person facing stigma is likely to withdraw from others, even more so if they feel ashamed, a situation that may lead to them disengaging from a service.

Therefore, it was vitally important that the guidelines addressed stigma and shame in three ways: first, to challenge the stigmatising attitudes that the reader may have; second, to offer guidance on how the reader could convey to a bereaved person that they were not stigmatising them; and third, to counter any stigma that a bereaved person may assume about the reader, whether the reader is actually stigmatising them or not.

The guidelines

As described earlier, the guidelines focus on five key messages. Each message used the following structure:

- Brief introductory paragraphs to explain the importance and relevance of the key message.
- Quotes taken from the research to illustrate the positive and negative impact that professionals and workers can have on bereaved people. These demonstrate to readers the importance of the key message, the difference their work can make and the importance of empathy.
  For key messages 1, 2 and 3 there is a question directed at the reader's own personal experience and inviting them to empathise with the bereaved person. For key messages 4 and 5 there is a question for the reader about their own working practice, inviting them to reflect on how they respond to bereaved people.

- Good practice guidance, from the simplest, most important and generic to the more complex and potentially specialised.

We now discuss each message.

Key message 1 – show kindness and compassion

This first message came out of powerful examples from interviewees of how a professional or worker showing kindness and compassion could be an antidote to any stigma the bereaved person might have experienced. It included being more available and giving of their time, and to convey respect for the bereaved person's privacy.

Working group members contributed several simple, realistic and practical ideas about how the reader could respond with compassion: for example, offering condolences; switching off mobile phones, radios and other communication devices; and just being there and listening. In addition, the reader is asked to consider '[W]hat it would be like for you to go and talk to a worker or professional about what you believe is a difficult and shameful thing in your life. What would that be like?'

The message therefore invites and encourages the reader to make 'human' contact with the bereaved person, whilst at the same time performing their role.

Key message 2 – language is important

Underlying the guidance about the language the reader uses in their work is the importance of breaking down stigma and shame. Use of language is also a significant part of how to put into practice Key message 1. Two quotes from bereaved people, which accompany key message 2, illustrate the impact of respectful and disrespectful language:

> And it is just a horrible stigma, you get a label on you, you are labelled . . . it is as if when she died 'Oh another one bites the dust'. That's the impression we got you know when it all happened, it was just horrible.
>
> (Mother talking about son)

> The doctor in A&E who signed his death . . . he said "This gentleman had died" and that made such a difference to us. We were upset and I thought he wasn't referred to as "This drug addict has died", you know.
>
> (Parent talking about son)

The guidance recognises how language varies across workplaces, offering suggestions not only for what to say but also *how* to say it.

This key message was informed by not only the experiences of working group members, reviewers and bereaved people, but also by other research evidence (Kelly and Westerhoff, 2009) about how language does matter when talking about substance use.

Key message 3 – every bereaved person is an individual

This message uses several themes from the interviews to help the reader to respond more sensitively to people bereaved by substance use. Interviewees stressed the importance of professionals responding to them as the person they were, that is, to their unique situation and often many and complex needs, rather than as just another death and/or bereavement. In some cases, this included looking beyond the cause of death and assumptions about addiction. Other research themes informing this key message include stigma, being available and taking time and respecting privacy.

Because this message is more about the reader's *attitude* to bereaved people rather than their behaviour, there is emphasis on explaining the wide range of responses bereaved people may receive, rather than specific guidance. This reflects working group members' appreciation of the diversity of bereaved people they had worked with and of treating people as individuals at a time when they are particularly vulnerable, for example, considering each member of a family and recognising different religious or cultural beliefs.

Key message 4 – everyone can make a contribution

The idea underlying this message is to encourage the reader to take personal responsibility for what they can actually do to help. This is achieved by asking the reader to take the same approach that is used to safeguarding children or vulnerable adults, where everyone has a responsibility to do what they can (Safeguarding Vulnerable Groups Act, 2006), familiar to many readers. Next readers are invited to reflect on their own work with bereaved people and then to consider the following contrasting experiences:

> The fact of the matter is I did feel desperately let down . . . a young man fighting for his life and surely somebody in that A&E Department would have had the decency to say, "Well I think he needs next of kin" but no, they didn't.
>
> (Mother talking about son)

> The police were great; they couldn't have been any nicer. They really couldn't. The liaison officer I think it was, there was a man and a woman and the two of them they really were awful, awful nice. They couldn't do enough to help us.
>
> (Mother talking about son)

The importance of this key message was clear from the interviews with bereaved people which identified the importance of professionals and workers being available and taking time; keeping the bereaved person informed and offering them help; and the need for continuity in support.

Again working group members' experience informed the good practice guidance, typified by not being afraid to talk about the death, countering a common tendency to avoid talking about difficult and painful subjects like bereavement.

Key message 5 – working together

The essence of this message is the importance of joint working between agencies to ensure both continuity of support and that bereaved peoples' complex and multiple needs are met.

The idea of joint working between agencies is familiar. However, the working group members found that in reality joint working is often difficult and inconsistent. Therefore, the reader is invited to consider their own work with bereaved people, followed by simple practical steps to improve joint working. For example, identifying and recording local and national services available to bereaved people would enable them to be easily 'sign posted' to those services.

Given the significance of both information and stigma the guidance pays particular attention to *how* to refer someone onto another service so as to maximise the likelihood of the referral actually being taken up.

Other significant sections

In addition to the five key messages, the guidelines included sections on 'What is bereavement?' and 'What are substance use and addiction?' These provide basic information for readers who do not have specialist knowledge and, in particular, challenge commonly held assumptions that are inaccurate and potentially unhelpful. These include, for example, that bereavement is something we 'get over' and that addiction is a lifestyle choice or the result of being weak-willed.

Another section entitled 'When bereavement and substance use come together' gives an overview of substance-related deaths to convey how these typically lead to severe bereavements entailing high levels of need. This section is particularly relevant for those readers who have not experienced a significant bereavement themselves and may be unaware of just how overwhelming and long term these bereavements can be. The section also challenges the assumption that those who die through substance use are necessarily addicted. It lists the range of ways people can die through substance use, as well as the range of experience that bereaved people may have had of the person's substance use before the death and its legacy after the death.

Dissemination

This section describes how the guidelines were disseminated to a wide range of professionals and workers primarily in the UK, along with brief comments on two online resources available on how to use the guidelines. As part of the guidelines' official launch we wanted to disseminate them as widely as possible.

Therefore, we circulated them to all relevant national organisations as well as to additional national and local organisations known to the research team, working group members, the reviewers of the guidelines, and others who supported the project. The guidelines were disseminated in both hard copy and a PDF format on the University of Bath's Centre for Death and Society (CDAS) website.

Some individuals ensured comprehensive coverage in their areas. There was also a recognition of different organisational cultures, where those with a hierarchical culture would need a 'top down' approach to dissemination and other organisations would need a 'bottom up' approach driven by individual professionals and workers.

Following this initial dissemination, recipients forward the guidelines on to their networks of colleagues, other local agencies and professional bodies. We assume that this forwarding process has continued, though it is not something that we have been able to track. Therefore, we do not know the true scale of dissemination, particularly as it is probably still ongoing.

Online resources

The team developed two free online resources to help professionals and workers learn how to use the guidelines in their work. They are both PowerPoint presentations that are available for download, at no cost, on the CDAS website and can be used for continuing professional development. One presentation, 'Guidance for practitioners' is designed to train professionals and workers in how to use the Guidelines, either individually through independent learning or as part of a group training session. The other presentation, 'Introduction to research project and guidelines', is about the project itself, and includes two slides on the Guidelines. Both of these presentations have been used by all the Coroners Offices in Kent.

How are the guidelines being used?

Before describing how the guidelines are being used by key organisations we provide an example of how they are being used by a professional working directly with a bereaved family. It comes from a colleague of one of the authors of this chapter who works for an addiction service.

> *In the course of my work a lot of effort goes toward supporting family members affected by someone's substance misuse. To highlight support through a service delivering the 5-Step Method,[8] I ran a workshop for family members focusing on what can I do when my loved one won't stop?*
>
> *Several months after the workshop a mother who attended the session contacted me and asked for the opportunity for her and her husband to meet me again. Their daughter had died of liver failure and they were devastated. I agreed to meet them and just before I did I received a copy of the Guidelines. I was relieved to receive this*

resource. I had agreed to see this couple, yet with 20 years' experience of working in the field I knew this could be an emotionally charged session.

The Guidelines were extremely helpful. As expected, the level of the couple's sadness, anger, guilt, along with feeling that they had somehow failed to prevent the death were evident.

What I found most helpful were the key messages particularly the first two, covering kindness and compassion along with good practice and the use of language.

It helped me be prepared but most importantly helped the session be supportive to two family members who were obviously hurting. What suggested the session was helpful was the couple asked for another session with an inquest approaching. I want to say thank you to those behind the Guidelines. It is practical and supports practitioners when they are having these types of emotive conversations.

This example demonstrates a benefit of the guidelines that we did not anticipate. Almost certainly this individual had the skills needed to work with bereaved families and already understood the guidelines' five key messages. However, the guidelines were useful for increasing confidence in supporting bereaved people. This was picked up on by a member of the working group who noted that they can be used by workers and professionals to aid their own personal reflection on their practice, particularly if they believe that their interaction with a bereaved individual or family could be or was difficult. For example, professionals involved at the time of death or in the immediate aftermath, such as paramedics, GPs, the police and the coroner, although they may have shown compassion in dealing with the bereaved person, such an interaction can still be very difficult and leave them reflecting on how it went and whether it could have been better. In this case the guidelines can be seen as a tool for 'debriefing' and reassuring the reader that they have handled a difficult situation to the best of their ability.

A member of the working group noted that change in practice does not occur 'overnight' and although the phrase 'learning by one's mistakes' may be an over-generalisation, developing the skills to demonstrate good practice is often an incremental process in which the guidelines provide a useful resource.

One main theme identified through discussions with key organisations was the guidelines being used by those with direct contact with families bereaved through drug or alcohol use in a way that could be described as reactive rather than proactive. For example, it was highlighted by a member of the advisory group that the guidelines were being used within their organisation as a resource when addressing any bad practice that had been identified when working with bereaved people. In this kind of situation this agency would provide a copy of the guidelines to the organisation concerned with a view to tackling bad practice (particularly practice that reinforces bad practice).

Some potential benefits of the guidelines are that they were produced as part of an independent, impartial, government-funded research study and by a working group with representation from professions such as police, paramedics and coroners whose response, in some instances, bereaved people considered could have been better. A single-issue charity or voluntary-sector organisation could be seen as biased or too close to the issue, therefore their material may be more easily dismissed. Therefore, although the guidelines are more generic, they may be harder to ignore.

Although the guidelines complement the material already produced by organisations which support families affected by drug or alcohol use such as DrugFAM in England and SFAD in Scotland, they may have particular value for smaller organisations. One advisory group member who supported the dissemination of the guidelines noted their particular value to small, local or specialist support groups. There is always a balancing act between producing guidelines that are too specialist and would not be read by certain groups, or too generic so that they are only considered as additional material. In general the 'less is more' approach agreed by the working group appeared to have been vindicated. As an example some mental health professionals noted the usefulness of the guidelines for workers and professionals who do not have a direct role in dealing with bereavement but who are likely to come into contact with bereaved families. A main benefit for those secondary workers and professionals was that the guidelines' brief accessible format fitted the typical amount of time they have to spend with bereaved people.

This penetration into the wider workforce did, unfortunately, appear to depend on a small number of geographically local, or individuals working in, specialist agencies. Few key individuals within the working group 'championed' the guidelines, either due to their role or the existing networks which allowed for easy dissemination; however, most of the busy workers and professionals who generously gave their time to produce the guidelines were not able to actively promote or disseminate them. The working group identified one potential failing of the study to be the limited resources available for dissemination. In the future, it may be worthwhile building in the role of 'champion' or 'cultural architect' into any project with appropriately costed dissemination.

Conclusion

This chapter has described the third and final phase of the project to improve how services respond to people bereaved through a substance-related death. This phase drew on the findings from the project's first and second phases on both bereaved people's as well as practitioners' experiences to identify five key messages that any professional or worker could use. These were then successfully developed into the first set of good practice guidance for working with these bereaved people. Finally, we described the extensive dissemination and highlighted how the guidelines are being used by professionals and workers who support individuals bereaved through alcohol or drug use.

How the guidelines could be developed

The guidelines are open source, so users are free to develop them. Here are some possibilities:

1   Provide the five key messages in a visual summary for professionals and workers. This could be either as a 'desktop' background image for a computer screen or a postcard that was given to all workers in an organisation.
2   The guidelines could usefully be modified into a version for use by the family and friends of people bereaved through a substance-related death. The five key messages have been adapted for this purpose by Alcohol Concern as a part of their online memorial project for people bereaved through alcohol use.
3   Versions or 'appendices' could be produced for professional groups who have a particular role with this group of bereaved people (e.g., counsellors, coroners, police and journalists). These new versions could form part of changing attitudes within a particular work area, as with The Samaritans' Media Guidelines (2013).
4   Versions could be produced for both families and professionals who come into contact with bereaved children and young people.
5   Further guidance could be developed to support bereaved people from diverse backgrounds, who can often experience organisations as discriminatory. They include people who are lesbian, gay, bisexual or transgender and people from particular faith groups who have beliefs associated with certain substances, such as bereaved Muslim families.

Notes

1   Cruse is a national charity in England and Wales set up to offer free, confidential help to bereaved people.
2   Winston's Wish is the leading childhood bereavement charity in the United Kingdom.
3   Alcohol Concern is a national charity in England and Wales whose aim is to reduce alcohol-related harm.
4   Adfam is the national charity whose aim is to improve life for families affected by drugs and alcohol.
5   DrugFAM is a charity set up by a mother who lost a son through substance use that provides support to people struggling to cope with someone's addiction to drugs or alcohol.
6   Cruse Scotland is a Scottish charity that supports people who have experienced the loss of someone close.
7   SFAD is a Scottish charity that supports families affected by someone else's alcohol and drug use.
8   The 5-Step Method is a brief intervention for families who are affected by a close other's substance use – see, for example, Copello et. al. (2010).

References

Ballatt, J., and Campling, P. (2011). *Intelligent Kindness: Reforming the Culture of Healthcare*. London: Royal College of Psychiatrists.

Cartwright, P. (2015). *Bereaved Through Substance Use: Guidelines for Those Whose Work Brings Them Into Contact With Adults After a Drug or Alcohol Related Death*. Bath: University of Bath. Available at: http://rebrand.ly/bereavementguidelines

Cole-King, A., and Gilbert, P. (2011). Compassionate care: The theory and the reality. *Journal of Holistic Health Care*, 8(3), 29–37.

Copello, A., Templeton, L., Orford, J., and Velleman, R. (2010). The 5-step method: Principles and practice. *Drugs: Education, Prevention and Policy*, 17(S1), 86–99.

Cruse Bereavement Care and the Bereavement Services Association. (2014). *Bereavement Care Service Standards*. Cruse Bereavement Care and the Bereavement Services Association.

HM Government. (2006). *Safeguarding Vulnerable Groups Act*. London: Stationary Office.

Kelly, J. F., and Westerhoff, C. M. (2009). Does it matter how we refer to individuals with substance-related conditions? A randomized study of two commonly used terms. *International Journal of Drug Policy*, 21, 202–207.

Ministry of Justice. (2014). *A Guide to Coroners Services* [online]. London, Available at: www.gov.uk/government/uploads/system/uploads/attachment_data/file/363879/guide-to-coroner-service.pdf Accessed 17th April 2016.

The Samaritans. (2013). *Media Guidelines for reporting suicide* [online]. Ewell, Surrey, Available at: www.samaritans.org/sites/default/files/kcfinder/branches/branch-96/files/Samaritans%20Media%20Guidelines%20UK%202013%20ARTWORK%20v2%20web.pdf

Templeton, L., Ford, A., McKell, J., Valentine, C., Walter, T., Velleman, R., Bauld, L., Hay, G., and Hollywood, J. (2016). Bereavement through substance use: Findings from an interview study with adults in England and Scotland. *Addiction Research & Theory*. 24 (5), 341–354 doi 10.3109/16066359.2016.1153632

Walter, T., Ford, A., Templeton, L., Valentine, C., and Velleman, R. (2015). Compassion or stigma? How adults bereaved by alcohol or drugs experience services. *Health and Social Care in the Community*. doi 10.1111/hsc.12273

Chapter 8

Conclusion

Christine Valentine and Linda Bauld

This book has reported on a study that set out to address both the lack of understanding as well as of adequate support for adults bereaved by substance use. In particular, we found that the stigma of substance use was a major theme in our interviews, affecting those left behind as well as the deceased. Indeed, interviewees' recollections both confirmed and extended existing findings (see, e.g., Guy, 2007) on how such stigma may marginalise bereaved people who, rather than receiving sympathy and support for their loss, may instead attract disapproval, indifference and rejection from all directions. However, our study has also identified a growing awareness by some practitioners of what these bereaved people are coping with and a willingness to tackle this situation. With the benefit of these practitioners' experience and expertise, the study has been able to produce the first evidence- and practice-based guidelines to improve the response of a wide range of services.

Based on qualitative, research the study's findings have captured the commonalities, complexities and diversity of how people experience this kind of bereavement (Templeton et al., 2016). These findings are contained in seven stand-alone chapters, with varied authorship, each chapter addressing a key theme or area related to substance use bereavement together with variations on each theme. This concluding chapter draws together these various threads to provide an overview and assessment of the studies' achievements, including the relationship between the findings and the methods we used to obtain these. The chapter also considers how these methods enabled five key messages to be translated into guidelines that can be embedded in existing practice.

Part I described the experiences and needs of these bereaved adults as represented in the 100 interviews we conducted. Interviewing used an open-ended, conversational approach that encouraged interviewees to tell their story, while also attempting to cover specific key areas. These included the relationship with the deceased before they died, the nature of the deceased's addiction, how the person died and the impact of the death on those left behind, finding support and memory-making. The findings show how four key elements – the life, the death, the stigma and the memory – may combine to make this kind of bereavement particularly difficult to grieve and that diversity must be taken into account within these common elements.

Part II presented findings from six practitioner focus groups on how to address the shortcomings of existing service provision to better meet the needs of these bereaved people; and how these findings informed the task of an inter-professional working group to develop good practice guidelines. The use of focus groups enabled us to bring together a range of practitioners and, in advance of meetings, share some extracts from the interviews with them so as to bring home these bereaved people's experiences with greater immediacy. Indeed, the work carried out in Part II never lost sight of what we found in Part I; the reported experiences of the bereaved people we interviewed have informed all aspects of the study and ultimately its main output, the practice guidelines.

Part I: coping

The life

Many of our interviewees made clear associations between how difficult things were for them before the death and the impact of the death itself. Therefore, Templeton and Velleman, in Chapter 1, considered how published studies on living with a family member's or close friend's substance use might inform how people subsequently coped with bereavement. More specifically they considered the relevance to bereaved families of the 'stress, strain, coping, support' (SSCS) model of the predicament of families living with a member's substance use. Based on the five elements of stress, strain, understanding, coping and support, this model suggests that stress and strain are exacerbated or relieved depending both on how the family understands and copes with the substance use as well as the availability of support. The authors found these five elements to be relevant to what our bereaved interviewees told us about their experiences before death, suggesting that that the SSCS model can also shed light on the experiences of bereaved families as an inevitable extension of the difficulties they had already faced before the death.

The death

In considering the impact of the death itself, Ford et al., in Chapter 2, analysed interviewees' experiences of the death both in the immediate aftermath and further down the line. These include the often difficult and distressing circumstances of the death itself, the daunting, confusing and time-consuming experience of navigating the system through which these deaths were officially dealt with and the difficulty of finding support for a poorly understood bereavement involving deaths that often attract social stigma. In so doing, the chapter reports on new findings on the implications for these bereaved people of how they are treated by professionals and practitioners, particularly in the immediate aftermath of a death. At a time when those left behind are likely to be at their most vulnerable, the responses of those in official positions are likely to have a profound impact for better or worse and with long-lasting effects on their grieving. In highlighting the impact of poor responses from professionals and practitioners in the immediate aftermath of these deaths, these findings break new ground in understanding the predicament of those bereaved by substance use. They therefore provided important material for the focus group discussions and the working group's task.

Stigma

How stigma affects both those who use substances and those who subsequently mourn them formed the focus of Walter and Ford's Chapter 3. In discussing the functions of stigma in society, the authors suggest that the stigma of substance use deaths is linked to how these deaths not only challenge cultural norms but also remind us of our mortality. Thus it is not surprising that stigma was a major theme in what our interviewees told us, including both direct stigma via comments from others that were unkind, inappropriate and judgmental, and perceived or anticipated stigma, which could prevent the bereaved person seeking support. However the authors also found that not every substance-related death was stigmatised, at least not by everyone the interviewee encountered. Furthermore, a significant number of interviewees were able to exercise their own agency in choosing, for better or for worse, how to respond to stigma, and thus developed ways to manage, stand up to and survive it, some of which were positively creative (Valentine and Walter, 2015). In addition, there was a very clear willingness among the practitioners in our focus groups to develop non-stigmatising practice.

Memory

Given the stigma of a substance use lifestyle and the painful and disturbing memories of the deceased person's life and death, Chapter 4 by Valentine and Templeton considered how a life and death involving substance use may be remembered and memorialised. Interviewees' accounts of both public and more private forms of remembering revealed diversity and ingenuity in how they met such difficulties. In recalling the more public aspects of remembering, interviewees talked about the impact of the funeral, including how it was organised and what it was like to be there; and press reporting of the death and the inquest. Interviewees' also recalled more private aspects of remembering, such as not wanting or finding it too painful to remember, or struggling to salvage fond memories among a predominance of bad. Drawing on concepts of post-traumatic growth and continuing bonds, the chapter identifies and discusses the creative ways some people responded to their loss, including projects designed to raise awareness of substance use, challenge social stigma and support others in similar situations.

Diversity

One important finding is that, although there appeared to be core elements to our 106 interviewees' experiences of bereavement through substance use, there was also marked diversity. In Chapter 5, Templeton and Velleman considered possible differences across a number of dimensions: relationship – the six ways in which our participants were related to those who died (parent, child, spouse or partner, sibling, friend, and extended family member); treatment/recovery for the interviewees' own substance use problems; substance type, comparing deaths involving alcohol with those involving drugs; and geography, comparing interviewee experiences in England with those in Scotland. Their findings suggest that, in addition to the core difficulties associated with these bereavements (Chapters 1–4), support should cater for diversity and be more widely available for sub-groups who felt very strongly that there was insufficient support available for them, most notably siblings, friends and extended family members. There is also scope for offering more bereavement support as part of or alongside the packages of care available to those who are themselves in treatment or recovery for their own alcohol or drug use.

Part II: services

Issues for professionals

In Chapter 6, McKell et al. sketched the complexity and fragmentation of the terrain through which bereaved family members have to find their way after a substance related death and within which officials and professionals have to operate. The chapter goes on to identify the challenges that the professionals who comprised our focus groups identified in responding to those bereaved from substance-related death: identifying, liaising with and referring to other organisations; workplace priorities and culture that do not prioritise bereavement; gaps in professional capabilities and training; and how to tailor support group provision to each individual's varied needs.

Improving support

In Chapter 7, Cartwright et al. described how both the interviews and the focus groups informed the final phase of the project – to improve how services respond to adults bereaved following a substance-related death. First, the authors considered the need for such guidance, identifying three main factors: the severity of the bereavement, the lack of existing guidance and being in a position to provide evidence-based guidance. Second, they described how our research data and ideas from wider policy research were drawn together to identify five key messages that any practitioner whose work brought them into contact with substance use deaths could use to improve their response to those left behind. These messages formed the backbone of a set of practice guidelines (Cartwright, 2015). Third, they described how the guidelines were produced by a working group of experienced and diverse professionals and practitioners and reviewed by an ad hoc group of other individuals with relevant experience. Fourth, the authors discussed how the guidelines have been widely disseminated and used within the UK to date and what we know of how they are being used to improve support for those bereaved after a drug or alcohol-related death. They concluded the chapter by providing some suggestions for further developing the guidelines.

Regardless of its achievements, our study is in some respects also a starting point. It is the first sizeable qualitative study of this kind of bereavement, combining substance use and bereavement expertise and working with a range of practitioner groups. As a result the study has provided new evidence of how this kind of bereavement may be particular to and therefore differ from other bereavements and has applied this evidence to developing practice guidelines. Yet inevitably gaps remain, which call for further research. We were not able to examine the experiences and needs of black and minority ethnic groups, children and young people, those bereaved by deaths resulting from new psychoactive substances (NPS) or how bereaved people's needs may vary according to their socio-economic status. However, it is also important to acknowledge what the size of our sample did enable – that is, to identify a combination of factors that are unique to this kind of bereavement as well as the diversity both within and between each interviewee's recollections. As a result it became apparent that the support needs of this group could not be standardised and that the challenge was to develop guidelines that encouraged practitioners to provide person-centred support while at the same time appreciating the special difficulties these bereaved people face. By centralising the experiences of our bereaved interviewees the guidelines invite the practitioner reader to identify with the bereaved service user, while highlighting both the specific challenges these bereaved people face as well as the particularity and diversity of each person's experience.

References

Cartwright, P. (2015). *Bereaved Through Substance Use: Guidelines for Those Whose Work Brings Them Into Contact With Adults Bereaved After a Drug or Alcohol-Related Death*. Bath: University of Bath. Available at: http://rebrand.ly/bereavementguidelines

Guy, P., and Holloway, M. (2007). Drug-related deaths and the 'Special Deaths' of late modernity. *Sociology*, 41(1), 83–96.

Templeton, L., Valentine, C., McKell, J., Ford, A., Velleman, R., Walter, T., Hay, G., Bauld, L., and Hollywood, J. (2016). Bereavement through substance use: Findings from an interview study with adults from England and Scotland. *Addiction Research and Theory*, 24(5), 341–354.

Index